Cleaning Validation

A Pocket Guide for Engineers

©*Priscilla Browne*

ISBN-13: 978-1974544318

ISBN-10: 1974544311

Contents

Introduction	8
What Is Cleaning?	8
Why Clean Equipment and Products?	8
Verification and Validation	9
Definitions	10
Regulatory Requirements	14
FDA	15
EU GMP	19
ICH Q7	21
PIC/s	24
Validation Standards	25
Stages of Validation	25
Stage 1 — Process Design	26
Stage 2 — Process Qualification	27
Stage 3 — Continued Process Verification	27
Validation: General Principles and Practices	32
Cleaning Validation	32
Pre-requisites to Cleaning Validation	33
Execution	34
Validation Report	34
Clean-in-Place (CIP)	35
Visibly Clean	37
Soils and Their behaviour	37
Detergents	38
Validation Strategies	39
Summary	40
How Are Acceptance Levels Defined?	41

Historical Context of Limits	43
Uses of the Term Limit	43
PDA Technical Report No. 29	44
Calculation of MACO	45
MACO for Each Piece of Equipment	47
Coverage Testing	47
Cleaning Validation Protocol	49
PIC/S Guidance on Limits	51
Test Methods	52
ICH Q7 Validation of Analytical Methods	52
Definitions	53
Cleaning Process Design	55
Equipment Considerations	57
Cleaning Agent Approval	58
Critical Cleaning Parameters	58
Cleaning Pipes	59
Dead Legs	61
Connections and Tie-ins	62
Welding	62
Valves	64
Materials of Construction	64
Stainless Steel	65
Pressure Testing	66
Passivation	66
Passivation Process	68
Stainless Steel and Rouging	70
P&ID	72

Sampling	**83**
Direct Sampling	83
Rinse Sampling	85
Sources of Contaminants	86
Microbiological Sampling	87
Utilities	**95**
Introduction	95
Key Definitions	95
Compressed Air	98
Water Systems	99
Clean Steam	102
Case Study 1 - Production Line Switching	107
Case Study 2 - Coating Pan Swab Site Selection	110
Inspection of Cleaning Processes	114
Appendix I Precision Cleaning (Medical Devices)	123
Glossary	124

Introduction

What Is Cleaning?

Cleaning can be defined as the process of removing potential contaminants from process equipment and maintaining the condition of equipment so that it can be safely used for subsequent product manufacture. It is complicated by many different chemicals used to produce medicinal drug products and other chemical agents used in the manufacturing process or in the cleaning process.

Why Clean Equipment or Products?

Facilities can be multi-product facilities, i.e. the same equipment is shared over different products. However, dedicated facilities also require cleaning evaluation and strategies. Cleaning minimises the transfer (or carry-over) of one product into another product, to an acceptable level, by means of product residue.

Some product residues are considered so toxic/potent if carried-over to another product that they are required to be manufactured in a dedicated facility (e.g. penicillin).

Equipment: Clean-in-place, often abbreviated to CIP, allows equipment cleaning to occur with minimal disassembly of equipment. CIP programs allow different products using similar or different materials to be manufactured on the same equipment.

Products: Supplies of products and medical devices used by patients or healthcare professionals must be clean and free of contamination.

Regulatory Requirements: It is the aim of every manufacturer to provide safe and effective products for use by patients and end users such as doctors and nurses. Companies are granted licenses to supply markets with products based on regulatory compliance and product safety. Cleaning compliance is a key part of achieving a state of compliance and more importantly, supplying safe products.

Verification and Validation

Verification: Verification means confirmation by examination and provision of objective evidence that specified requirements have been fulfilled[1].

When it comes to cleaning, if the cleaning procedures have not been fully validated, the effectiveness of the cleaning procedure should be verified at the completion of cleaning. This is "verification".

Validation: Validation means confirmation by examination and provision of objective evidence that the particular requirements for a specific intended use can be consistently fulfilled[1].

---

[1] 21CFR820.3

Definitions

Clean Hold Time (CHT): The total time the parts or equipment are held clean post-cleaning.

Cleaning Agent: The chemical agent or solution used as an aid in the cleaning process.

Cleaning Process Parameters: The parameters that are critical in the cleaning process. Subsequent cleaning process monitoring may or may not utilise these parameters.

Critical Process Parameter (CPP): A control parameter that has a direct relationship to the quality, safety, effectiveness or performance of the intermediate or final product.

Dirty Hold Time (DHT): The total time the parts (or equipment) are held dirty prior to cleaning.

Maximum Allowable Carry Over (MACO): The amount of allowed product residue carry-over from lot-to-lot, batch-to-batch, etc. This limit is based on the lowest of:

(1) Limited based on toxicity,
(2) Limit based on smallest therapeutic dose, and
(3) Worst case dose methodology

Residue: Substances left on surfaces of equipment after cleaning that may pose a risk for subsequent use. Example: residues that may require cleaning include: products, excipients, raw materials/intermediates, non-volatile solvent, non-intrinsic cleaning agents such as detergents, etc.

Worst Case Conditions: Conditions considered to pose the greatest chance of process or product failure. The highest or lowest value of a given control parameter or set of parameters.

Visual Inspection: With regard to cleaning, visual inspection should be completed by appropriately trained and experienced personnel on completion of equipment/process clean-down.

Surfaces should be visibly clean and free of visible residue. Hard-to-clean places should be examined in particular.

cGMP: Current Good Manufacturing Practices

Concurrent Validation: Validation activities occurring at the same time as one another or concurrent to a product launch.

Prospective Validation: This is when validation is done in advance of commercial manufacturing.

Protocol: An approved document that contains the tests and verifications to be conducted during the validation. Validation protocols include test methods and test conditions, acceptance criteria and parameters required.

FAT (Factory Acceptance Test): Typically classified as an engineering activity, the purpose of the FAT is to verify that the equipment or system meets the requirements of the URS.

Deviation: An event which results in failure with respect to the acceptance criteria in the protocol.

Process window: The selected operating range of machine settings/parameters that will produce product to meet all quality and product specifications.

Installation Qualification: Establishing through documented evidence that all functionality of the process equipment meets the manufacturer's specification and company requirements.

Equipment Qualification: Providing confidence through documented evidence that the equipment is suitable for the intended use and is capable of consistently operating within set limits and tolerances.

Operational Qualification: Providing confidence through documented evidence that the product can be manufactured to specifications within set limits and tolerances.

(MVP) Master Validation Plan: A governing document which sets out the validation approach and provides details of deliverables. An MVP should be written as soon as possible and it should align and reflect with the "current" validation strategy.

Precision Cleaning Systems: A precision cleaning system is a piece of equipment that can remove soil or dirt from parts or components. Most precision cleaning systems are made up of several stages, e.g. clean-rinse-dry. The simplest cleaning systems consist of one stage, e.g. an ultrasonic bath containing heated water.

Clean-in-Place (CIP): CIP is a cleaning method used to clean the inner surfaces of piping, vessels and process equipment without the need for disassembly.

PIC/s: The Pharmaceutical Inspection Convention and Pharmaceutical Inspection Co-Operation Scheme (referred to as PIC/S) are two international bodies between countries and pharmaceutical inspection authorities that co-operative on subjects relating to the field of GMP.

Skid: This is essentially a modular process that can be plugged into a process onsite with little construction or integration required. Skids are used as part of clean-in-place solutions within the food and beverage industry as well as pharmaceutical industries.

Regulatory Requirements

Key regulatory and international publications are included below:

➤ FDA – Food and Drug Administration – Guide to Inspections of Validation of Cleaning Processes

➤ EU GMP – European Commission – EudraLex Volume 4: EU Guidelines to Good Manufacturing Practice, Medicinal Products for Human and Veterinary Use, and Annex 15 (section 10 "Cleaning Validation")

➤ ICH Q7 – International Council on Harmonisation - Good Manufacturing Practice

➤ Guide for Active Pharmaceutical Ingredients (section 12.7 "Cleaning Validation")

➤ ICH Q9 – International Council on Harmonisation – Quality Risk Management

➤ PIC/S PI 006-3 – Pharmaceutical Inspection Co-Operation Scheme – Recommendations on Validation Master Plan, Installation and Operational Qualification, Non-Sterile Process Validation, Cleaning Validation (section 7 "Cleaning Validation")

➤ WHO TRS 937 – World Health Organisation - Specifications for Pharmaceutical Preparations; Annex 4: Supplementary guidelines on Good Manufacturing Practices: Validation; Appendix 3: Cleaning Validation

FDA – Food and Drug Administration - Guide to Inspections of Validation of Cleaning Processes

Cleaning validation programmes are important requirements for both bulk pharmaceutical processing and biotechnology. As with validation of other processes, there may be more than one way to validate a cleaning process. Once the manufacturer can establish inspection consistency and repeatable outcomes that ensure predetermined acceptable criteria are met, a cleaning procedure can be deemed effective. This is a driven process which should support claims of consistent outcomes.

It isn't solely the FDA that has an expectation that cleaning procedures (processes) be validated; PIC/s, ICH, EudraLex and WHO guidance and requirements also specify the need to validate cleaning procedures.

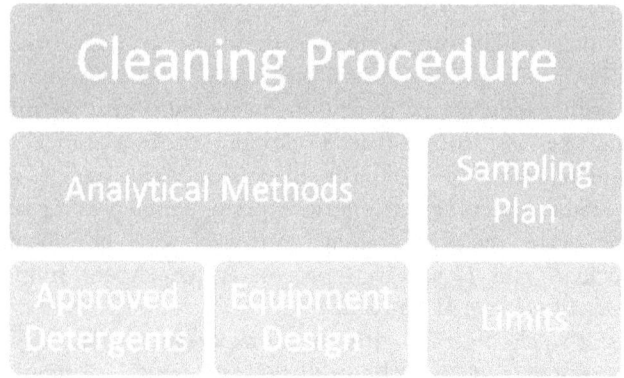

Figure 1: Components of a Cleaning Programme

A Historical Lesson

Historically, the FDA was mostly concerned about the contamination of non-penicillin drug products with penicillin or the cross-contamination of drug products with potent hormones or steroids. One event which increased FDA awareness of the potential for cross-contamination due to inadequate procedures was the 1988 recall of a finished drug product, Cholestyramine resin USP.

In this instance, the bulk pharmaceutical used to produce the product had become contaminated with low levels of both intermediates and degradants.

The cross-contamination in this case was attributed to the reuse of recovered solvents. The recovered solvents had been contaminated because of a lack of control over the reuse of solvent drums. Drums that had been used to store recovered solvents from a pesticide production process were later used to store recovered solvents used for the resin manufacturing process. Some shipments of this pesticide-contaminated bulk pharmaceutical were supplied to a second facility at a different location for finishing. This resulted in the contamination of the bags used in that facility's fluid bed dryers with pesticide contamination.

General Requirements:

Upon inspection by auditors, the following requirements are expected in order to demonstrate a robust and suitably validated cleaning programme (procedure):

➢ Written procedures (SOPs) detailing the cleaning processes used.

➢ Register or list of dedicated equipment. Fluid bed dryer bags are another example of equipment that is difficult to clean and are often dedicated to a specific product.

➢ A written procedure on how cleaning processes are validated.

➢ Written validation protocols detailing the sampling procedures and analytical methods to be used including the sensitivity of those methods and acceptance criteria.

➢ A validation report which is approved in advance of commercial manufacturing. The data generated during the validation should demonstrate that residues have been reduced to an "acceptable level".

Evaluation of Cleaning Validation:

The main focus of an auditor in respect of cleaning validation is to evaluate the evidence that aims to demonstrate the effectiveness of the approach and processes used to clean equipment.

The following questions are relevant when evaluating the cleaning process:

- At what point does a piece of equipment /system become clean?
    - This knowledge should be captured in cycle development and development of the cleaning process. Studies may indicate that a vessel or piece of equipment requires three rinses with hot water-for-injection at which point it meets acceptance criteria. However, an additional number or rinses may be included to provide a level of confidence in the cleaning process.

- Does it have to be scrubbed by hand?
    - Depending on the drug substances and excipients or other chemicals, residues may tend to physically "stick" to surfaces or behave as tarry or gummy which may require mechanical force to remove them, or a solvent rinse may be sufficient for removal.

When the cleaning process is used only between batches of the same product, a company may only meet the criteria of "visibly clean" for the equipment. This can often be referred to as a batch-to-batch clean. Such between-batch cleaning processes do not require validation. Change-over from one product to a different product of different materials requires a more comprehensive clean, potentially requiring multiple cleans or rinses.

EU GMP – European Commission – EudraLex Volume 4: EU Guidelines to Good Manufacturing Practice, Medicinal Products for Human and Veterinary Use, and Annex 15 (section 10 "Cleaning Validation")

Section 10 of Annex 15 provides a number of bullet points with regard to cleaning validation:

"Cleaning validation should be performed in order to confirm the effectiveness of any cleaning procedure for all product contact equipment. Simulating agents may be used with appropriate scientific justification. Where similar types of equipment are grouped together, a justification of the specific equipment selected for cleaning validation is expected.

A visual check for cleanliness is an important part of the acceptance criteria for cleaning validation. It is not generally acceptable for this criterion alone to be used. Repeated cleaning and retesting until acceptable residue results are obtained is not considered an acceptable approach.

It is recognised that a cleaning validation programme may take some time to complete and validation with verification after each batch may be required for some products, e.g. investigational medicinal products. There should be sufficient data from the verification to support a conclusion that the equipment is clean and available for further use.

Validation should consider the level of automation in the cleaning process. Where an automatic process is used, the specified normal operating range of the utilities and equipment should be validated.

For all cleaning processes an assessment should be performed to determine the variable factors which influence cleaning effectiveness and performance, e.g. operators, the level of detail in procedures such as rinsing times etc. If variable factors have been identified, the worst case situations should be used as the basis for cleaning validation studies.

Limits for the carryover of product residues should be based on a toxicological evaluation. The justification for the selected limits should be documented in a risk assessment which includes all the supporting references. Limits should be established for the removal of any cleaning agents used. Acceptance criteria should consider the potential cumulative effect of multiple items of equipment in the process equipment train.

The risk presented by microbial and endotoxin contamination should be considered during the development of cleaning validation protocols.

The influence of the time between manufacture and cleaning and the time between cleaning and use should be taken into account to define dirty and clean hold times for the cleaning process.

Where campaign manufacture is carried out, the impact on the ease of cleaning at the end of the campaign should be considered and the maximum length of a campaign (in time and/or number of batches) should be the basis for cleaning validation exercises.

Where a worst case product approach is used as a cleaning validation model, a scientific rationale should be provided for the selection of the worst case product and the impact of new products to the site assessed. Criteria for determining the worst case may include solubility, cleanability, toxicity and potency.

Cleaning validation protocols should specify or reference the locations to be sampled, the rationale for the selection of these locations and define the acceptance criteria.

Sampling should be carried out by swabbing and/or rinsing or by other means depending on the production equipment. The sampling materials and method should not influence the result. Recovery should be shown to be possible from all product contact materials sampled in the equipment with all the sampling methods used.

The cleaning procedure should be performed an appropriate number of times based on a risk assessment and meet the acceptance criteria in order to prove that the cleaning method is validated.

Where a cleaning process is ineffective or is not appropriate for some equipment, dedicated equipment or other appropriate measures should be used for each product as indicated in chapters 3 and 5 of EudraLex, Volume 4, Part I.

Where manual cleaning of equipment is performed, it is especially important that the effectiveness of the manual process should be confirmed at a justified frequency."

Ref: EU GMP V4, Annex 15, 2017

ICH Q7 – International Council on Harmonisation - Good Manufacturing Practice:

"Cleaning procedures should normally be validated. In general, cleaning validation should be directed to situations or process steps where contamination or carryover of materials poses the greatest risk to API quality. For example, in early production it may be unnecessary to validate equipment cleaning procedures where residues are removed by subsequent purification steps. (12.70)

Validation of cleaning procedures should reflect actual equipment usage patterns. If various APIs or intermediates are manufactured in the same equipment and the equipment is cleaned by the same process, a representative intermediate or API can be selected for cleaning validation. This selection should be based on the solubility and difficulty of cleaning and the calculation of residue limits based on potency, toxicity, and stability. (12.71)

The cleaning validation protocol should describe the equipment to be cleaned, procedures, materials, acceptable cleaning levels, parameters to be monitored and controlled, and analytical methods. The protocol should also indicate the type of samples to be obtained and how they are collected and labelled. (12.72)

Sampling should include swabbing, rinsing, or alternative methods (e.g., direct extraction), as appropriate, to detect both insoluble and soluble residues. The sampling methods used should be capable of quantitatively measuring levels of residues remaining on the equipment surfaces after cleaning. Swab sampling may be impractical when product contact surfaces are not easily accessible due to equipment design and/or process limitations (e.g., inner surfaces of hoses, transfer pipes, reactor tanks with small ports or handling toxic materials, and small intricate equipment such as micronisers and microfluidisers). (12.73)

Validated analytical methods having sensitivity to detect residues or contaminants should be used. The detection limit for each analytical method should be sufficiently sensitive to detect the established acceptable level of the residue or contaminant. The method's attainable recovery level should be established. Residue limits should be practical, achievable, verifiable, and based on the most deleterious residue. Limits can be established based on the minimum known pharmacological, toxicological, or physiological activity of the API or its most deleterious component. (12.74)

Equipment cleaning/sanitation studies should address microbiological and endotoxin contamination for those processes where there is a need to reduce total microbiological count or endotoxins in the API, or other processes where such contamination could be of concern (e.g., non-sterile APIs used to manufacture sterile products). (12.75)

Cleaning procedures should be monitored at appropriate intervals after validation to ensure that these procedures are effective when used during routine production. Equipment cleanliness can be monitored by analytical testing and visual examination, where feasible. Visual inspection can allow detection of gross contamination concentrated in small areas that could otherwise go undetected by sampling and/or analysis. (12.76)"

PIC/S PI 006-3 – Pharmaceutical Inspection Co-Operation Scheme – Recommendations on Validation Master Plan, Installation and Operational Qualification, Non-Sterile Process Validation, Cleaning Validation (section 7 "Cleaning Validation")

PIC/s provides several pages of recommendations on cleaning validation. It clearly outlines the principles and purpose of conducting cleaning validation:

"Pharmaceutical products and active pharmaceutical ingredients (APIs) can be contaminated by other pharmaceutical products or APIs, by cleaning agents, by micro-organisms or by other material (e.g. air-borne particles, dust, lubricants, raw materials, intermediates, auxiliaries). In many cases, the same equipment may be used for processing different products. To avoid contamination of the following pharmaceutical product, adequate cleaning procedures are essential.

Cleaning procedures must strictly follow carefully established and validated methods of execution. This applies equally to the manufacture of pharmaceutical products and active pharmaceutical ingredients (APIs). In any case, manufacturing processes have to be designed and carried out in a way that contamination is reduced to an acceptable level.

Cleaning Validation is documented evidence that an approved cleaning procedure will provide equipment which is suitable for processing of pharmaceutical products or active pharmaceutical ingredients (APIs).

The objective of cleaning validation is confirmation of a reliable cleaning procedure so that the analytical monitoring may be omitted or reduced to a minimum in the routine phase."

Validation Standards

➢ ASTM E2281-03 "Standard Practice for Process and Measurement Capability Indices".

➢ ASTM E2500-07 "Standard Guide for Specification, Design, and Verification of Pharmaceutical and Biopharmaceutical Manufacturing Systems and Equipment".

➢ ASTM E2709-09 "Standard Practice for Demonstrating Capability to Comply with a Lot Acceptance Procedure".

Stages of Validation

This section provides a background on general principals in respect of validation. Process validation can be divided into three stages which are:

➢ Process design
➢ Process validation
➢ Continued process monitoring

Similarly, cleaning validation can be divided into the same three stages; cleaning process design, cleaning process validation and continued process monitoring of cleaning processes.

Prior to the commercial manufacture and distribution of drug products and medicines to consumers, a manufacturer must gain a high degree of assurance in the performance of the manufacturing process such that it will consistently produce APIs and drug products meeting those attributes relating to identity, strength, quality, purity and potency. A high degree of assurance and demonstrated consistency of a process must be evidence-based.

In order to best prepare a process that is consistent and produces safe and effective products, the following points must be understood and implemented:

- ➤ Understand the sources of variation
- ➤ Detect the presence and degree of variation
- ➤ Understand the impact of variation on the process and ultimately on product attributes
- ➤ Control the variation in a manner commensurate with the risk it represents to the process and product

<u>Stage 1 — Process Design</u>

The goal of this stage is to design a process suitable for routine commercial manufacturing that can consistently deliver a product that meets its quality attributes.

The activity as this point is focused on defining the commercial manufacturing process that will be used going forward and that the records and associated documents support the shaping of the commercial process.

Designing an efficient process with an effective process control approach is dependent on the process knowledge and understanding of personnel. This is achieved through pilot studies and testing according to sound scientific methods and principles, including good documentation practices.

ICH Q10 Pharmaceutical Quality System also highlights the importance of process knowledge and the extent of its capability.

It is also a regulatory requirement to establish a strategy for process control. Strategies for process control can be designed to:

- Reduce input variation
- Adjust for input variation during the manufacturing process
- Take into account a combination of reduction and adjustment

Having a method of process control reduces variability and gives a higher assurance of consistent product quality.

Stage 2 — Process Qualification

During the process qualification (PQ) stage of process validation, the process design is evaluated to determine if it is capable of reproducible commercial manufacture.

This stage has two elements:

(1) Design of the facility and qualification of the equipment and utilities

(2) Process performance qualification (PPQ)

Stage 3 — Continued Process Verification

The goal of stage three, continued process validation, is to provide assurance that the process remains in a state of control (the validated state) during commercial manufacture.

Systems for detecting unplanned departures from the process as designed are essential to accomplish this goal. Adherence to the cGMP requirements, specifically, the collection and evaluation of information and data about the performance of the process, will allow detection of undesired process variability.

Evaluating the performance of the process identifies problems and determines whether action must be taken to correct, anticipate and/or prevent problems so that the process remains in control (§ 211.180(e)).

An ongoing programme to collect and analyse product and process data that relate to product quality must be established (§ 211.180(e)). The data collected should include relevant process trends and quality of incoming materials or components, in-process material and finished products.

The data should be statistically trended and reviewed by trained personnel. The information collected should verify that the quality attributes are being appropriately controlled throughout the process.

The FDA recommends that a statistician or person with adequate training in statistical process control techniques develop the data collection plan and statistical methods and procedures used in measuring and evaluating process stability and process capability.

Procedures should describe how trending and calculations are to be performed and should guard against overreaction to individual events as well as against failure to detect unintended process variability. Production data should be collected to evaluate process stability and capability. The quality unit should review this information.

If properly carried out, these efforts can identify variability in the process and/or signal potential process improvements. Good process design and development should anticipate significant sources of variability and establish appropriate detection, control and/or mitigation strategies as well as appropriate alert and action limits.

Key FDA Recommendations:

- The use of quantitative, statistical methods whenever appropriate and feasible. Scrutiny of intra-batch as well as inter-batch variation is part of a comprehensive continued process verification programme under § 211.180(e).
- Continued monitoring and sampling of process parameters and quality attributes at the level established during the process qualification stage until sufficient data are available to generate significant variability estimates.
- Process variability should be periodically assessed and monitoring adjusted accordingly. Variation can also be detected by the timely assessment of defect complaints, out-of-specification findings, process deviation reports, process yield variations, batch records, incoming raw material records and adverse event reports.
- Production line operators and quality unit staff should be encouraged to provide feedback on process performance.
- The quality unit meets periodically with production staff to evaluate data, discuss possible trends or undesirable process variation and coordinate any correction or follow-up actions by production.
- Maintenance of the facility, utilities and equipment is another important aspect of ensuring that a process remains in control. Once established, qualification status must be maintained through routine monitoring, maintenance, and calibration procedures and schedules.

Reference: FDA Process Validation: General Principles and Practices January 2011 — Current Good Manufacturing Practices (CGMP) Revision 1

Some references that may be useful in understanding the general principles of process validation (which are also relevant to cleaning validation) include the following:

- ASTM E2281-03 "Standard Practice for Process and Measurement Capability Indices,"

- ASTM E2500-07 "Standard Guide for Specification, Design, and Verification of Pharmaceutical and Biopharmaceutical Manufacturing Systems and Equipment,"

- ASTM E2709-09 "Standard Practice for Demonstrating Capability to Comply with a Lot Acceptance Procedure."

Note: This is not a complete list of all useful references on this topic.

Validation: General Principles and Practices

Validation has a number of common requirements that apply to equipment qualification, process validation or cleaning validation.

- ➢ Documented evidence (records/reports)
- ➢ Demonstration of a high degree of assurance (data-driven studies and data analysis)
- ➢ Consistency (traditionally three PQ runs)
- ➢ Predetermined quality attributes (product specifications)

Cleaning Validation

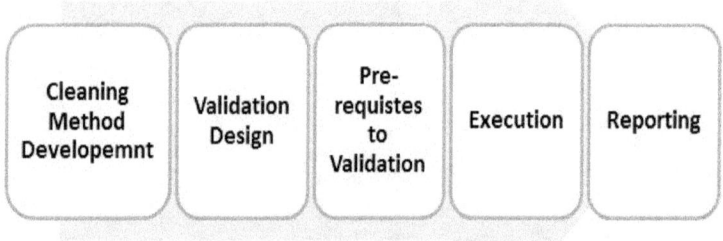

Figure 2: Life Cycle of Cleaning Validation

The purpose of cleaning validation is to provide objective evidence that methods and procedures are capable of removing product residues, contaminants, cleaning agents used, by-products, solvents and degradants to below a predetermined level.

All contamination is referred to as soiling. A simple method of examining sources of contamination is reviewing the input materials (product ingredients, manufacturing agents etc.) of a process.

Risk assessment must be applied to decide on the extent of cleaning validation studies.

Risk assessment based on product exposure is required to determine the need for validation of facility cleaning procedures (e.g. floors, walls).

Cleaning validation must demonstrate the following with documented evidence (reports and records):

➢ the effectiveness of the cleaning procedure against contamination (chemical or microbiological)

➢ effectiveness of the cleaning procedure against product cross-contamination

➢ control of the critical parameters (times, concentration, temperature)

Prerequisites to Cleaning Validation

Procedures must specify the level of cleaning to be undertaken, cleaning intervals and frequency and the methodology to be utilised. The procedures must be well defined to ensure consistency of operation whether they are manual or automated.

Cleaning validation protocols must describe the test methods to be used (e.g. residue test methods) and limits, microbiological limits and scientific rationales for those limits.

Physio-chemical and microbiological test methods used for cleaning validation must be validated and capable of detecting residues to the required level. Calculations based on scientifically justified limits usually result in impractically high values. For microbial tests, the limit of 25 CFU per cm2 (25 CFU/cm2) is typically used for non-sterile manufacture.

As appropriate, development of cleaning procedures must be performed and documented. Worst-case considerations must be applied. Sampling techniques must be defined and authorised. Rationales for using these techniques must be available. An approved sampling plan must be in place.

Execution

Typically, three consecutive applications of the cleaning procedure must be performed and shown to be successful in order to prove a cleaning procedure is validated.

A maximum holding period between end of use and cleaning should be evaluated as part of the validation study/plan. (Dirty hold time.)

A maximum holding period between cleaning and re-use must also be evaluated to determine how long equipment, facilities or systems may remain idle following cleaning. (Clean hold time.)

Validation Report

The results of the cleaning validation should be presented in an approved cleaning validation report, clearly stating the outcome. Discussion of any issues or non-conformances along with resolutions should also be included.

<u>Clean-In-Place (CIP)</u>

Cleaning validation for a CIP system design involves the intersection of two similar or different products. Take a simple example: a pharmaceutical company manufactures two types of paracetamol caplets (tablets).

Product A contains the active ingredient paracetamol, preservatives and other excipients.

Product B is also a paracetamol product but it contains an additional ingredient, caffeine. Therefore, product B is branded differently and marketed with a more discerning customer in mind.

Where multiple products are manufactured on the same equipment or machinery, the process is often referred to as non-dedicated.

As with the above example, if the same equipment is used to produce product A and Product B, an intersection of products occurs.

Product A- Cleaning must be effective enough to remove residues to acceptable levels.

Product B- When manufacturing commences, the residue levels must not contaminate the product.

Residue is any substance or trace of substance left on equipment or surfaces after cleaning. It is near possible to remove all residues from surfaces, so a residue limit should be medically safe and at a level that does not cause product quality issues or concerns.

Visibly Clean

Within any cGMP environment, the requirement to maintain a clean and suitable manufacturing area is key to compliance and ensuring product quality and customer safety. Visual inspection of the cleaning process must be done before swabbing. Inspection should confirm the equipment is visually clean and dry and no adverse odours are present.

Upon completion of visual inspection, swabbs should then only be taken if required by procedure. For areas that cannot be accessed for visual inspection or swabbing, a rinse sample can be taken in place of a swab.

Sometimes it is not possible to obtain a swab or rinse sample, therefore visual inspection may be the only method used to verify cleaning effectiveness.

In any validation, an important theme is to challenge the consistency of a process. Therefore, samples must be representative to ensure a proper picture is painted. Sampling sites should be taken from "hard-to-clean" areas as well as "easy-to-clean" areas to ensure that samples are representative of the equipment.

Soils and Their Behaviour

Soils are a source of contamination to products and therefore can present a risk to patients or users. Soils can be introduced by unplanned and unintended events, but they are likely a part of the process, or as a result of a manufacturing agent that has been used within a manufacturing process. Examples include coolant of cutting fluid used in a machining process. The fluid is required to achieve a good surface finish and reduce tool wear. The presence of this soil of parts can potentially be:

- ➢ Dried on during subsequent process step
- ➢ Compacted
- ➢ Dried on during dirty hold time
- ➢ Baked on during an oven process

Detergents

Cleaning agents and solvents must be sourced and approved according to a supplier qualification procedure. The following information should be considered for inclusion in a supplier qualification file.

- ➢ Certification to a quality management system such as FDA, ISO
- ➢ Supplier questionnaire
- ➢ Product specification
- ➢ Material safety data sheets
- ➢ Change notification policy
- ➢ Expiration dating (format and controls)
- ➢ Onsite audit
- ➢ Statements of suitability
- ➢ Sample certificate of analysis

Acceptance criteria for cleaning agent residues should be based on the lowest **LD50** of each chemical in an agent's formulation.

The standard LD50 cleaning agent's classification applies to all agents which contain chemical components whose LD50 is greater than 100 mg/kg. For low LD50 cleaning agents, the classification applies to all agents which contain chemical components whose LD50 is less than or equal to 100 mg/kg.

Validation Strategies

In this section we examine validation strategies. There are two main approaches for consideration:

(1) Direct approach

The direct approach consists of validating the cleaning procedure for all pieces of equipment and for all the products made.

(2) Matrix approach (aka grouping approach, family approach, bracketing)

A family or matrix approach can also be used where similar products can be grouped together with a representative.

A matrix can be formed and justified by defining a set of parameters and characteristics so that limited numbers of parameters or quality attributes are representative of the group. It should also focus on the worst-case parameters and quality attributes. All of the information that provides a rationale for implementing a matrix approach should be documented with a risk assessment created.

The following criteria may be considered to define the worst-case product with regard to cleaning validation:

- PDE
- Solubility
- Cleanability

With regard to medical devices, a particular product size of product configuration may be selected to represent the worst-case product. Therefore, by qualifying the worst case, all other products within the family of products would be considered validated.

With regard to pharmaceutical products, e.g. solid dose manufacturing of pain killers, products of a similar chemistry/content can be grouped together.

Summary

Grouping/matrix approach can be done by:

- product (soil)
- equipment
- worst case

Advantages of grouping/matrix approaches:

- Potential to simplify the amount of validation work
- Fewer validation runs

Establishing Residue Limits and Acceptance Criteria

Typical target residues on product within precision cleaning systems include:

- ➢ Organic residuals
- ➢ Particulate
- ➢ Bioburden

> Endotoxin

Typical target residues for CIP systems include:

> Drug active
> Cleaning agent
> Bioburden
> Endotoxin
> Degradation products or by-products

How Are Acceptance Levels Defined?

Several considerations need to be accounted for when establishing safe and effective residue levels.

– Consider the potential effects of target residue on subsequent products or raw materials
– Pharmacology of residue
– Toxicity of residue
– Stability issues

As per European guidance, (Reference Human Drug CGMP Notes, 9:2, 2Q 2001) equipment does not have to be as clean as the best possible method of residue detection or quantification, as absolute cleanliness is required or feasible. However, it should be as clean as can reasonably be achieved – "to a residue limit that is medically safe and that causes no product quality concerns….."[2]

---

[2] Human Drug CGMP Notes, 9:2, 2Q 2001)

Historical Context of Limits

Pre-1993 Industry Acceptance Limits:

- $1/10^{th}$ of therapeutic dose
- $1/50^{th}$ of max therapeutic dose
- Less than smallest therapeutic dose
- 3ppm (arsenic)
- 30ppm for cleaning agents
- Detection limit of method

This approach was inconsistent from company to company, arbitrary and not based on risk.

1993 Eli Lilly Article:

Gary Fourman and Dr. Mike Mullin in "Pharmaceutical Technology" April 1993 proposed:

- $1/1000^{th}$ of a dose in max daily dose
- 10ppm of product in another product (next product)
- No residue visible

Uses of the Term "Limit"

L0 = Daily amount allowed per patient (μg or mg) (L zero)

This has been called the Acceptable Daily Intake (ADI), Acceptable Daily Exposure (ADE), Permitted Daily Exposure (PDE). The limit is based on safety and toxicity information. Typical values used for L0 include: 0.001 of minimum daily dose of active ingredient based on toxicity information.

L=1 Concentration in next product: PIC/s guidance suggests a maximum of 10ppm.

L=2 Absolute amount in manufacturing vessel train (mg): [MAC – maximum allowable carryover] – L2

This limit uses the absolute amount in manufacturing vessels. It is calculated by multiplying L1, limit, times the batch size of subsequent product to be manufactured.

L3=Amount per surface area ($\mu g/cm^2$): This is calculated by dividing L2 by shared surface area of the equipment train (the sum of surfaces).

L4a = Amount per swab (μg): The amount per swab depends on the limit per surface area; (L3) – swabbed area.

Calculate:

L4a = L3 X Swabbed Area

L4b Conc. in swab extraction solution:

Concentration in swab extract depends on:

- Limit per surface area (L3)
- Swabbed area
- Amount of solvent present for extraction

Calculate:

L3 X Swabbed Area/solvent extraction amount.

L4c = Conc. in "rinse" water (µg/g): L4b change based on volume for extraction

If sample is extracted into 10g of solvent:

100 µg / 10 g = 10 µg/g

NOTE: L=0, Daily amount allowed is also known as:

Acceptable Daily intake (ADI)

Acceptable Daily Exposure (ADE)

Permitted Daily Exposure (PDE)

Safe Daily Intake (SDI)

Values for L0 can be a minimum daily dose of active 0.001 or a value based on toxicity data (MSDS sheets etc.)

PDA Technical Report No. 29

The PDA technical report proposes the following limits:

- 1/10th -1/100th for topicals
- 1/100th – 1/1,000th for oral products
- 1/1,000th – 1/10,000th injections & opthalmics
- 1/10,000th – 1/100,000th for research or investigational products

Calculation of MACO (Maximum Allowable Carryover)

There are two steps in calculating MACO residue levels.

Firstly, it is necessary to **calculate the MACO from one batch to the next batch.**
The second step is to calculate the **allowable "drug" or "residue" of each unique product contact surface, for each piece of equipment.**

This then provides a **calculation based on the overall equipment train** (aka the equipment line).

The **MACO** is based on three calculations, which are:

- **MACO** based on toxicity
- **MACO** based on the smallest therapeutic drug dose
- **MACO** based on smallest solution batch size

$$NOEL = \frac{LD50 \times NHW}{2000}$$

$$ADI = \frac{NOEL}{SF}$$

$$MAC0 = \frac{ADI \times SSBS}{LNDD}$$

NOEL = No observed limit effect
LD50 = Lethal dose of drug
NHW = Nominal human weight
2000 = Is a constant factor for calculating NOEL
ADI = Allowable daily intake
SF = Safety factor e.g. 1000
SSBS = Smallest solution batch size

MACO based on smallest therapeutic drug dose (STDD):

$$Product\ Carry\ Over = \frac{STDD}{SF}$$
$$MAC0 = Product\ Carry\ Over \times \frac{SSBS}{LNDD}$$

MACO based on smallest solution batch size (SSBS):

$$Worst\ Case\ Number\ of\ Doses = \frac{SSBS}{LNDD}$$
$$MACO = LNDD \times Worst\ Case\ Number\ of\ \frac{Doses}{SF}$$

Using the above calculations, **the MACO for the equipment train can be determined.**
The MACO for each individual piece of **equipment of surface can then be calculated.**

MACO for Each Piece of Equipment

To calculate the MACO (allowable residue for each piece of equipment) you will need to have the following information available:

- **MACO** (per calculations above).
- **Surface area** of each piece of equipment and the total of the equipment train.

Coverage Testing

Process equipment often contains critical surfaces that need to be cleaned according to validated procedures. Examples include mixing vessels and freeze-dryer chambers. The removal of any residual contamination from these surfaces needs to be demonstrated. This is typically done with an easily detectable tracer such a riboflavin (e.g. for simple visual detection).

It is important to make the distinction between (1) coverage testing of equipment and (2) coverage testing of components that are subject to cleaning. However, the principle is the same in both instances.

For equipment, coverage testing should be performed as part of equipment qualification for all process-contacting surfaces. Coverage testing verifies that all process-contacting surfaces are wetted by cleaning liquids and identifies any potential blind spots or hard-to-clean locations on the equipment.

Locations on equipment that are not adequately cleaned can be identified through riboflavin fluorescence testing using UV light inspection.

Regulatory requirements do not specify a requirement for spray coverage testing. However, as per US Code of Federal Regulations and EudraLex Volume 4 Part II, equipment should be of appropriate design to facilitate cleaning.

The Pharmaceutical Inspection Convention and Pharmaceutical Inspection Cooperation Scheme (PIC/S) specify "critical areas (i.e. those hardest to clean) should be identified, particularly in large systems that employ semi-automatic or fully-automatic clean-in-place (CIP) systems."

A 2004 FDA warning letter included two separate mentions of inadequate spray ball coverage: "Your firm failed to establish and follow written procedures to ensure the cleaning and maintenance of equipment used in the manufacture, processing, packing or holding of a drug product 21 CFR 211.67(b) and 600.11(b).

For example, cleaning validation for the clean-in-place (CIP) process vessel which is utilised in the aseptic formulation of trivalent bulk influenza vaccine, did not include an assessment of the spray ball coverage for the vessel. The spray ball is used for cleaning product contact equipment.... In addition, the cleaning validation did not include an assessment of the spray ball coverage for the tanks."

Ref: FDA Warning letter issued to Chiron corporation, December 9, 2004)

<u>Examples of Riboflavin Solution Strengths</u>

250mg/1 litre of water (1:4 ratio), generally suitable for spiking components.

2g per 1 litre (1:5),
200mg/L (0.2g/L) solution of riboflavin, suitable for testing interior surfaces of tanks/vessels.

1g of riboflavin in 10 Litres of water 100ppm solution, suitable for coverage testing of equipment surfaces.

Cleaning Validation Protocol

The validation protocol is a formal document that is preapproved prior to its use of execution. It defines the prerequisites, methods, specific requirements, activities and acceptance criteria for the cleaning validation at hand.

The protocol should address the following:

- ➤ Scope and Objectives
- ➤ Approval by cross-functional team
- ➤ Responsibilities
- ➤ Signature and training log
- ➤ Equipment/area to be cleaned under study
- ➤ Critical cleaning parameters
- ➤ Sampling methods and sample plan
- ➤ Maximum hold times
- ➤ Acceptance criteria

> Number of cycles

The dirty hold time (DHT) aka dirty equipment hold time (DEHT) is the time lapse between the end of manufacture and the start of cleaning. The purpose of this time control is to limit the difficulty of the cleaning before residues or remaining products are allowed to dry out.

The clean hold time (CHT) aka clean equipment holding time (CEHT) is the maximum time equipment can sit idle before a re-clean is required prior to its use. The main purpose of this time control is to limit microbial contamination.

The drying time (DT) is another time control that aims to limit microbial growth within vessels or equipment. It is important to dry equipment immediately after it has been cleaned.

PIC/S Guidance on Limits

The Pharmaceutical Inspection Convention and Pharmaceutical Inspection Co-operation Scheme (jointly referred to as PIC/S) are two international instruments between countries and pharmaceutical inspection authorities which provide together an active and constructive co-operation in the field of GMP.[3]

The most important point to remember when it comes to limits is that residues meet predefined criteria, the most stringent criteria as listed below:

(a) No more than 0.1% of the normal therapeutic dose of any product should appear in the maximum daily dose of the following (next) product,
(b) No more than 10 (parts per million, ppm) of any product will appear in another product, (this value is not always the default)
(c) No quantity of residue should be visible on the equipment after cleaning procedures are completed. Spiking studies should determine the concentration at which most active ingredients are visible. [4]

The method of determining residue limits of active ingredients is based on an approach developed by Fourman and Mullen (1993) and is referenced in PIC/s guidance amongst other publications.

---

[3] http://www.picscheme.org/

[4] http://www.picscheme.org/publication.php?id=4

Test Methods

It is important to consider test methods and test method validation early on in the validation life cycle. A test method is a process or an action used to verify that a product feature or particular requirement meets a predefined specification. Test methods can be physical or analytical in nature. Test method validation should be completed in advance of cleaning as the test method is used to verify the outputs of such cleaning validations.

ICH, Q7, Validation of Analytical Methods

"Analytical methods should be validated unless the method employed is included in the relevant pharmacopeia or other recognised standard reference. The suitability of all testing methods used should nonetheless be verified under actual conditions of use and documented. (12.80)

Methods should be validated to include consideration of characteristics included within the ICH guidance on validation of analytical methods. The degree of analytical validation performed should reflect the purpose of the analysis and the stage of the API production process. (12.81)

Appropriate qualification of analytical equipment should be considered before initiating validation of analytical methods. (12.82)

Complete records should be maintained of any modification of a validated analytical method. Such records should include the reason for the modification and appropriate data to verify that the modification produces results that are as accurate and reliable as the established method. (12.83)

(Reference: Q7 Good Manufacturing Practice Guidance for Active Pharmaceutical Ingredients Guidance for Industry September 2016.)

Test method validation must address the following parameters in order to demonstrate suitability to the intended use of the method and ensure it is capable of achieving consistent performance.

Definitions

Specificity: The ability to assess unequivocally the analyte in the presence of components, which may be expected to be present. These can include impurities, degradants etc. (ICH Q2)

Linearity: The ability of an analytical procedure (within a known range) to obtain test results that are directly proportional to the concentration of analyte in the sample. (ICH Q2)

Range: The interval between the upper and lower concentrations (amounts) of analyte in the sample (including these concentrations) for which it has been demonstrated that the analytical procedure has a suitable level of precision, accuracy and linearity. (ICH Q2)

Accuracy: Expression of closeness of agreement between the value which is accepted either as a conventional true value or an accepted reference value and the value found. (ICH Q2)

Robustness: A measure of the capability of an analytical procedure to remain unaffected by small, but deliberate variations in method parameters and which provides an indication of its reliability during normal usage (ICH Q2)

Precision: Expression of the closeness of agreement of an analytical procedure (degree of scatter) between a series of measurements obtained from multiple sampling of the same homogeneous sample under the prescribed conditions. Precision may be considered at three levels: repeatability, intermediate precision and reproducibility. Repeatability expresses the precision under the same operating conditions over a short interval of time. Repeatability is also termed intra-assay precision. Intermediate precision expresses within-laboratories variations: different days, different analysts, different equipment, etc. Reproducibility expresses the precision between laboratories (collaborative studies, usually applied to standardisation of methodology). ICH Q2)

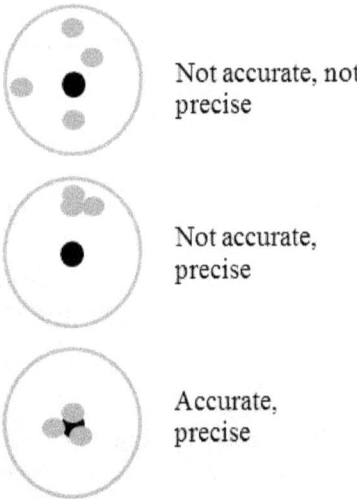

Figure 3: Illustration of accurate versus precise.

Detection Limit (LOD): The lowest amount of analyte in a sample that can be detected but not necessarily quantitated as an exact value for an individual analyte procedure (ICH Q2)

Quantitation Limit (LOQ): The lowest amount of analyte in a sample which can be quantitatively determined with suitable precision and accuracy for an analytical procedure. The quantitation limit is a parameter of quantitative assays for low levels of compounds in sample matrices and is used particularly for the determination of impurities and degradation products (ICH Q2).

Total organic carbon (TOC) analysis is a fast and effective analytical test method used for cleaning verification and validation in pharmaceutical manufacturing. It is used to test for residues of previously manufactured products (actives and excipients), cleaning detergents, chemicals, solvents, degradants and microbial contaminants. Detergent selection is a critical step in the development a cleaning programme. The purpose of the detergent is to remove residues; however, detergent levels remaining post-cleaning are undesired. If detergents remain post-cleaning they can affect the result of analytical tests.

Cleaning Process Design

For clean-in-place (CIP), the key elements to be considered in the design stage include:

– Equipment to be cleaned

– Soils to be removed

– Cleaning methods

– Cleaning agents

– Cleaning mechanisms

– Cleaning parameters

– Residue limits

CIP recipes should include the following information and parameters at a minimum:

- type of water (DI, USP purified, WFI) for pre-rinse,
- wash, and post-rinse
- volumes, times, flow rates, and pressures
- pump speeds
- process air blow times
- cleaning agent identification, concentration
- fill volumes to achieve circuit volume/flow
- alarm set points for parameters being monitored
- temperature and conductivity (rinses and cleaning solutions)

With regard to precision cleaning of medical devices, the design inputs are similar to clean in place. However, the focus of the cleaning is on the product that is processed, not the equipment itself.

– Product material and product type to be cleaned

– Soils to be removed (e.g. greases, oils)

– Cleaning methods (solvent or aqueous-based systems)

– Cleaning agents (detergents, nitric acid)

– Cleaning mechanisms (ultrasonic, heat, agitation)

– Cleaning parameters (temperatures, times, ultrasonic frequency)

– Residue limits (acceptance criteria)

Equipment Considerations

Firstly, for precision cleaning systems, the choice of equipment must be based on the intended purpose of the equipment, for example, will it be used for intermediate cleaning or for final cleaning? As you may expect, there are more stringent acceptance criteria for cleanliness when it comes to final cleaning precision equipment.

For CIP systems, the intended purpose of the equipment is a key design consideration. Materials of construction should be in keeping with the process, maintenance and cleaning regimes associated with it. Material certification should be provided by the supplier or vendor to ensure the correct grade of materials is used from approved suppliers.

In summary, the design should take into account:

- Difficult-to-clean locations
- What legacy systems are in place (hence knowledge)?
- Materials of construction cleaning agents to be used
- Cleaning parameters
- Individual cleaning versus cleaning as an equipment train

Cleaning Agent Approval

(a) Precision Cleaning Equipment

For aqueous-based systems, detergents are the preferred cleaning agents used. Detergents are water-soluble cleaning agents that "stick" or cause soils and dirt to "bind" together.

Detergents are typically diluted with de-ionised water or suitably clean water. Dosing can range from 5%-10%, though the dose depends on the type of soils present and the equipment. Ultrasonics, temperature and agitation provide for a quicker cleaning cycle.

(b) Clean-in-Place (CIP)

Cleaning agents used with CIP should address the following points:

- Effectiveness with regard to API/material
- Rinsability
- Supplier certification and compliance
- Stability
- Toxicity
- Supply and global availability
- EHS compliance

Critical Cleaning Parameters

The key parameters for cleaning can be remembered by using the acronym TACCT.

Time

Action
Cleaning chemistry
Concentration
Temperature

Mixing/flow
Water quality
Rinsing

Other cleaning parameters include flow rate, consideration of turbulence, water quality and rate of rinsing.

Critical cleaning parameters should be challenged at lower end (least stringent) of the operating window conditions during validation.

In cleaning development and design studies, parameters can be pushed to the point of failure of below the anticipated operating ranges. By completing more in-depth testing in the design phase, process knowledge is gained. It also can allow the cleaning validation to be more focused and avoid a protracted validation and may cut down on the number of cycles required if the evidence is there to risk assess a family or matrix validation approach.

Example: Temperature set-point is expected to be 80 ± 5°C.
– Perform cleaning development studies at set-point of 75°C (this covers the expected range and accounts for calibration tolerances).

<u>Cleaning Pipes</u>

The efficiency of a cleaning process is influenced by the type of flow within the system; the two types of flow are laminar flow or turbulent flow.

Laminar flow is when fluid particles move in parallel layers, at a constant velocity.

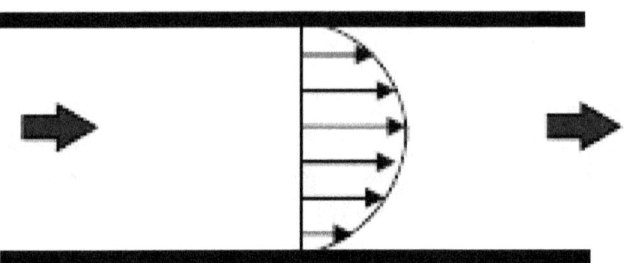

Turbulent flow is when the movement of fluid particles varies in velocity and direction.

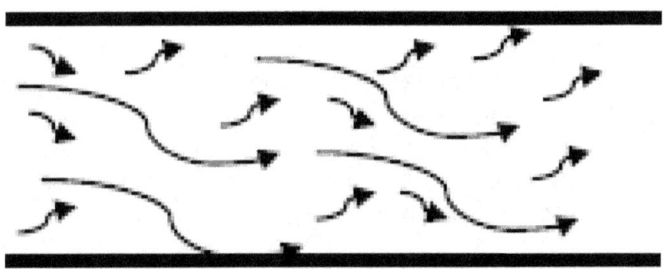

The Reynolds number of a system determines if the flow is turbulent or laminar. A Reynolds number (Re) greater than 4000 is described as turbulent flow.

$$Re = \frac{316Q}{dK}$$

Q= volumetric flow, (gal/min)

d= internal diameter (inches)

k= viscosity (cP)

<u>Dead Legs</u>

A dead leg is the world of piping terminology refers to an area of piping where there is insufficient flow or a tendency for water build-up or stagnation. The formal definition of a dead leg states that pipelines for the transmission of purified water for manufacturing or final rinse should not have an unused portion greater in length than 6 diameters (6D rule) of the unused portion of pipe measured from the axis of the pipe in use.[5]

---

[5] FDA guidance

Connections and Tie-Ins

Precision cleaning systems and CIP systems both have the necessity to be connected to utility supplies such as process and de-ionised water. Precision cleanliness will require tie-in of water supply and drainage on a continuous or defined frequency. Welding is the preferred permanent method of connecting pipes. Non-permanent connections are also used which allow the disconnection and swap-out of piping, vessels and equipment. Orbital welding is a common method of welding when joining piping assemblies and vessels. *(see next section)*.

Welding

Stainless steel process piping can be orbitally welded. Quality inspection is typically done real-time by designated quality inspectors using a fibrescope.

Welding should meet necessary standards such as the visual weld criteria as detailed in the materials joining part of the ASME BPE-2000 standard. Discolouration of the weld can be evident as a result of the high degree of heat. The discolouration is a result of the oxidation and can result reduce the corrosion resistance of the weld. In general, welds should not exhibit cracks, crevices or other surface deformities or visual defects. In the event of a weld failure, this can lead to system contamination and would result in the system not being in compliance with 21 CFR 211(a).

Figure 4: Acceptable weld penetration

Figure 5: Mismatch or misaligned weld

Figure 6: Outer Diameter concavity

Figure 7: Inner diameter concavity, aka suckback

Figure 8: insufficient penetration

Valves

Clamp-type connections can also be used for non-permanent connections. With regard to the use of valves, electromechanical valves that can be PLC controlled are preferred to manual valves. The level of automation depends on the complexity of the system. For example, a precision clean line used to clean metallic hip implants may have five or six clean and rinse tanks, all fitted with inlets and inlet valves. Having automated control is essential to running a complex line safely and efficiently

Materials of Construction

When it comes to materials of construction, the same selection criteria can be applied to precision cleaning systems and CIP equipment trains. Above all, materials and their surfaces should be non-reactive, non-corrosive and non-porous. Stainless steel of a high grade is often the preferred material of construction. Examples of grades used include 304, 316 and 316L.

For surfaces that are product contacting, material certificates are required to provide evidence that the materials and their constituents are of the correct make-up and suitable grade.

Stainless Steels

Stainless steels (SS) are crystallised solutions of at least 11% of corrosion reducing elements like chromium and nickel in iron. Generally, they are iron based with 12 to 30% chromium, 0 to 22% nickel and minor or no amounts of carbon, columbium, copper, manganese, molybdenum, nitrogen, phosphorus, selenium, silicon, sulfur, tantalum, and titanium.

Casting grades generally are designed with more sulphur to facilitate welding and have more ferrite (a less corrosion resistant phase) to prevent the formation of micro-cracks on cooling.

Preferred stainless steels for use in the life sciences are manufactured by VIM – vacuum-induction-melt followed by a secondary VAR – vacuum-arc-re-melt process with sulphur add-back and dispersion in order to minimise inclusions (stringers) and control the amount of sulphur used.

The surface of stainless steel can also be contaminated with the electrolyte solution used in electro-polishing if it is used an excessive amount of times or if rinsing steps are not adequate.

This solution builds up iron and other contaminants that can be transferred to the part being electro-polished if the conditions and chemistry are not carefully controlled.

To prevent these problems from occurring, all electrolyte solutions must be removed from the surface by using a chemically pure water rinse until the conductivity of the rinse from the stainless steel is equal to the conductivity of the water being supplied for rinsing.

Pressure Testing

Piping or system integration can be required for:

> Precision cleaning systems where the utilities need to be "tied" in to the system

> Installation of a pharmaceutical process within a facility e.g. a skid[6] plug in.

After installation, (and before passivation if required) piping systems are pressure tested by filling the system with clean air to 150% of the design pressure or 150psi, whichever is the greatest value. The pressure is then monitored over a 4-hour period to see if there is any drop in pressure.

Passivation

Passivation can be described as the active chemical process used to obtain a uniform chromium oxide layer on stainless steel (SS) surfaces. The chromium oxide layer or film forms a protective coating that gives corrosion resistant properties.

The protective layer naturally forms from the reaction of oxygen in air with the chromium on the metal surface but this naturally forming layer can be non-uniform or patchy due to impurities and surface chemistry defects.

---

6 see definition in introductory pages

When stainless steel is worked such as in welding, machining, mechanical polishing etc., its uniformity of the naturally forming protective layer can be damaged and oxides of other compounds forming the stainless steel composition can occur. Corrosion can begin at these non-uniform sites and, because stainless steel contains over 60% iron, the corrosion can proliferate from the surface through the body of the metal if no opportunity for protective layer reformation is given.

There are three passivation processes that are used to enhance the corrosion resistance of stainless steel:

- ➢ Treatment with oxidising acids
- ➢ Treatment with chelants
- ➢ Treatment by electro-polishing

Layer formation is a dynamic chemical process where the chromium atoms are combined with oxygen (and hydroxyl ions in aqueous environments) to form a complex surface layer that prevents attack on corrosion-prone atoms such as iron. Nickel and molybdenum may play a role in formation of the passive film but the mechanism has not been proven.

Passivation Process

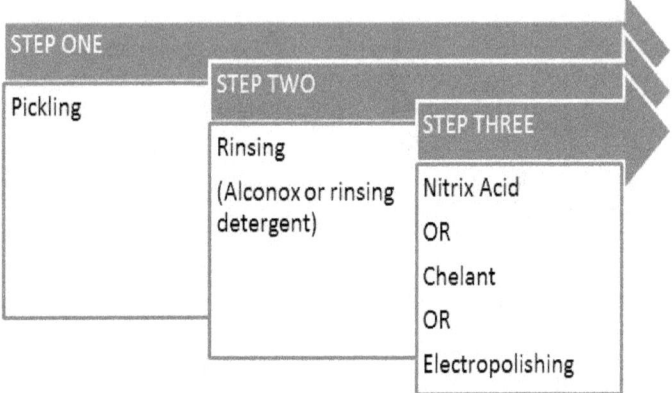

Figure 9: Passivation - three step process.

Passivation must start with a surface free of any oxide scale (including heat tints and oxide corrosion products) and organic contamination (machining lubricants, oils, coolant and grease).

The first step in a successful passivation process is pickling the metal at the mill. Pickling is a complete surface element and impurity removal process using a mixture of concentrated hydrofluoric and nitric acid.

The second passivation step occurs after fabrication and is the removal of organic contamination by washing of the surface with a basic trisodium phosphate (TSP) such as Alconox or other commercially available chelants-containing, free-rinsing detergent at elevated temperatures.

After the organics are removed, there are three commonly used processes used to complete passivation. The first uses a hot mineral acid solution (commonly nitric acid). The second method uses chelants with milder organic acids (citric acid) and sequestrants. The third is electro-polishing.

Mineral Acid:

This is a fast and affordable method of passivation; however, it comes with environmental and safety risks.

Nitric acid is the acid of choice because of its oxidising properties. The solution is usually heated to facilitate the oxidation reactions. However, the concentration has to be kept below 20% due to the metal surface removal that occurs at higher concentrations. Ph, temperature and conductivity of the acid are also monitored during the process.

Chelant:

Chelant processes are chemical in nature and the materials and their concentrations used can be adjusted to target particular contaminants or all likely residues on the metal.

The published data shows that chelant passivation can achieve chromium enrichment on the surface of stainless steel down to a depth of approximately 20 angstroms. This is a much higher enrichment and a greater depth of penetration than can be achieved by mineral acid passivation processes.

Chelant processes are proprietary but have the following points in common:

> - Heating of solutions to assist the chemical kinetics of the processes. The passivation solution is usually heated.
>
> - Mild organic acid to oxidise the surface iron to soluble ferric ions and insoluble ferrous ions.

> One or more chelants and sequestering agents to capture the ferric and ferrous ions and prevent their precipitation or depositing on the surface of metal.

Electro-polishing:

Electro-polishing uses a conductive, aqueous salt, reducing acid bath using sulfate and phosphate salts and acids along with a substantial direct current power source to remove 0.1 to 2.5 mils of surface metal preferentially from the peaks and high points. It can also remove surface inclusions, free the surface from iron and nickel, carbon and other surface contaminants to a maximum depth of approximately angstroms. This removal will enrich the surface of the stainless steel in chromium and therefore a highly passive surface is developed.

The article to be electro-polished is suspended in the conductive liquid and connected to the anode of the power supply. A second electrode is also suspended in the conductive liquid and is connected to the cathode. In order to achieve an even metal removal and chromium enrichment, it is important to achieve constant electrical potential across the surface of the article. Electro-polishing is limited to improving the surface evenness by approximately 10 Ra.

Stainless Steel and Rouging

Rouging is a form of corrosion found in stainless steels. It can be due to iron contamination of the stainless steel surface due to welding of non-stainless steel for support columns, or other temporary means, which when welded off leaves a low chromium area.

There are three classes of rouging:

> Class I – the stainless steel surface and the Cr/Fe ratio of the metal surface beneath such deposits usually remain unaltered.

> Class II - iron particles originating in-situ on unpassivated or improperly passivated stainless steel surfaces. By their formation, the Cr/Fe ratio of the metal surface is altered.

> Class III - iron oxide (or scale) which forms on surfaces in high temperature steam systems. The Cr/Fe ratio of the protective film is usually altered.

(Ref: ASME-BPE)

References

> EN 2516:1997 – Passivation of Corrosion Resisting Steels and Decontamination of Nickel Base Alloys

> ASTM A380 – Practice for Cleaning, Descaling and Passivating of Stainless Steel Parts, Equipment and Systems

> ASTM A967 – Specification for Chemical Passivation Treatments for Stainless Steel Parts

> ICH Q3D – International Council for Harmonisation – Guidance for Elemental Impurities

P&ID

Introduction

A piping and instrumentation diagram (P&ID) can be defined as:

1. A diagram which shows the interconnection of process equipment and the instrumentation used to control the process. In the process industry, a standard set of symbols is used to prepare drawings of processes. The instrument symbols are typically based on ISO 10628 International Society of Automation (ISA) Standard (S5.1).
2. The primary schematic drawing used for laying out a process control installation.

They usually contain the following information:

- Process piping, sizes and identification, including:
- Pipe classes or piping line numbers
- Flow directions
- Interconnections references
- Permanent start-up, flush and bypass lines
- tag identifiers),
- Valves and their identifications (e.g. isolation, shutoff, relief and safety valves)
- Control inputs and outputs (sensors and final elements, interlocks)

P&IDs are originally drawn up at the design stage from a combination of process flow sheet data, the mechanical process equipment design and the instrumentation engineering design. During the design stage, the diagram also provides the basis for the development of system control schemes, allowing for further safety and operational investigations, such as a hazard and operability study (HAZOP). To do this, it is critical to demonstrate the physical sequence of equipment and systems, as well as how these systems connect.

The most important symbols in relation to cleaning validation include understanding the vessels, tie-ins, valves and instrumentation associated with the equipment under validation.

A valve regulates, directs, or controls the flow of a fluid by opening, closing, or partially obstructing passageways in a piping system. This category includes orifices and other types of valves. The following pages provide a non-exhaustive list of some examples of common symbols.

Above: Gate valve, back pressure valve and needle valve.

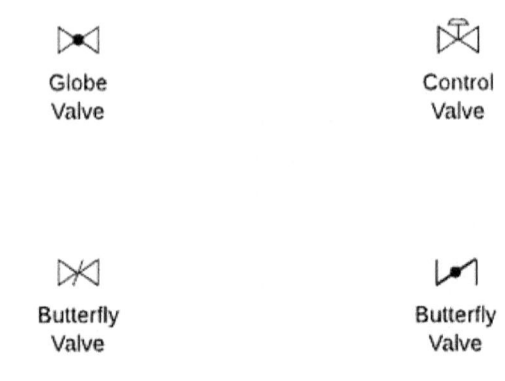

Above: Globe valve, butterfly valve and control valve.

Above: Ball valve, diaghragm, globe valve and check valve.

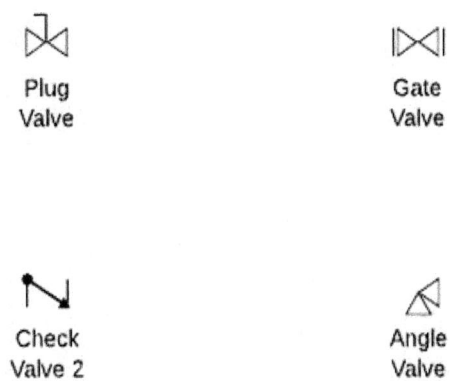

Above: Plug valve, gate valve, check valve two and angle valve.

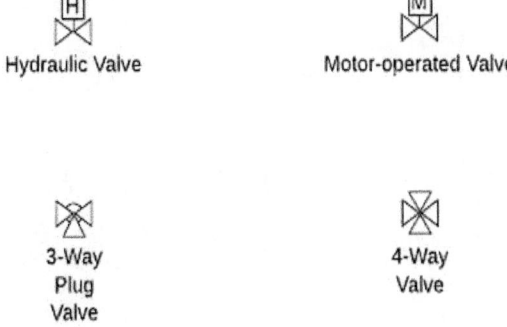

Above: Hydraulic valve, motor operated valve, three-way plug valve and four-way plug valve.

Powered Valve

Solenoid Valve

Float-operated Valve

Needle Valve

Above: Powered valve, solenoid valve, float-operated valve and needle valve.

Vessels

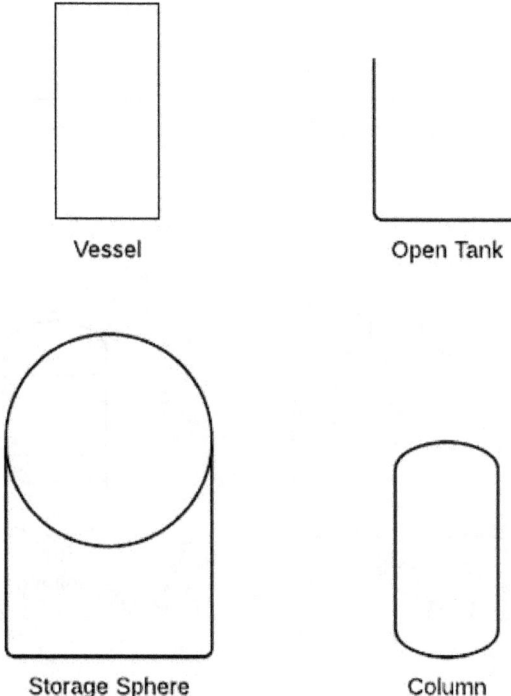

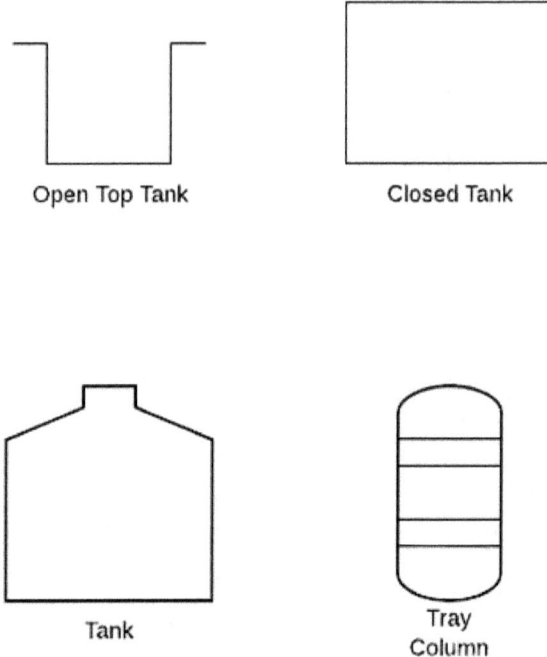

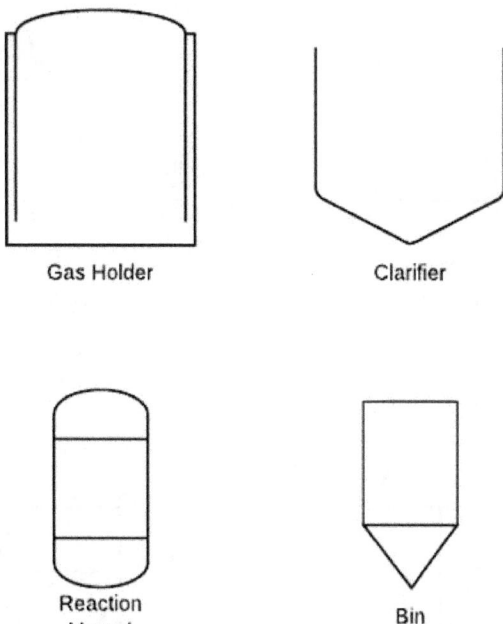

Instruments

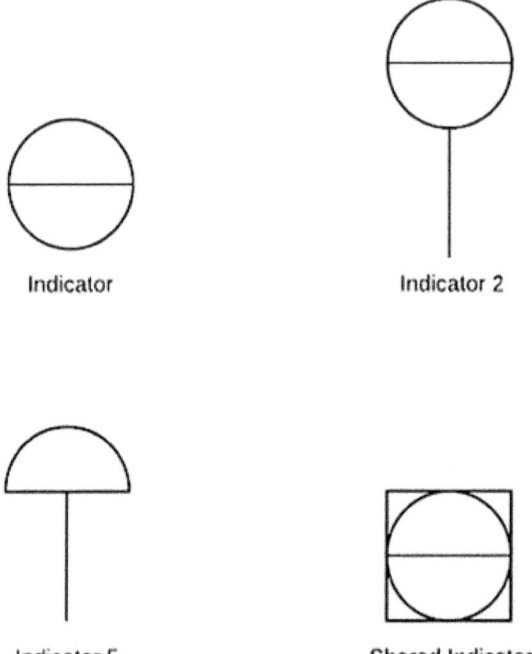

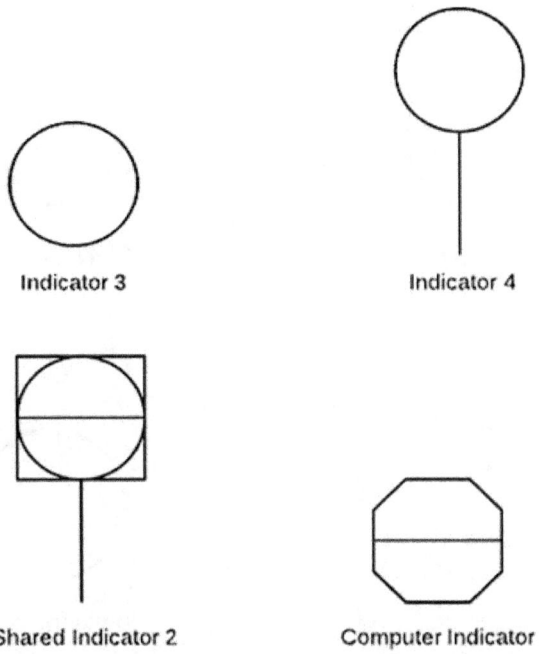

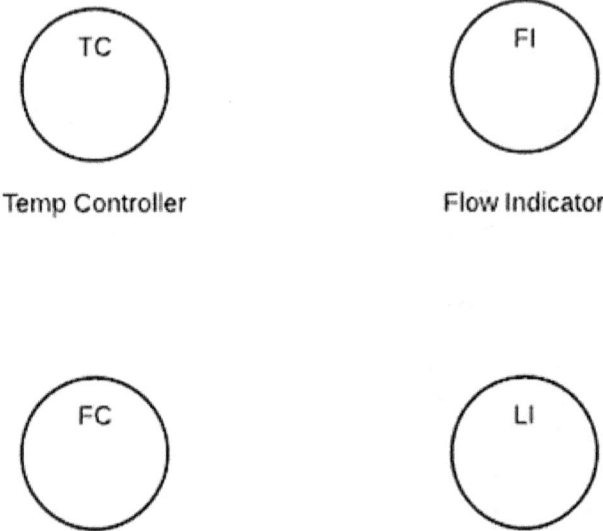

Sampling

There are two main methods of sampling that are considered to be acceptable; direct surface sampling and indirect sampling (use of rinse solutions).

A combination of the two methods is generally the most desirable, particularly in circumstances where accessibility of equipment parts can mitigate against direct surface sampling.

Direct Surface Sampling

(i) The suitability of the material to be used for sampling and of the sampling medium should be determined. The ability to recover samples accurately may be affected by the choice of sampling material. It is important to ensure that the sampling medium and solvent are satisfactory and can be readily used. Ref: PIC/S PI 006-3.

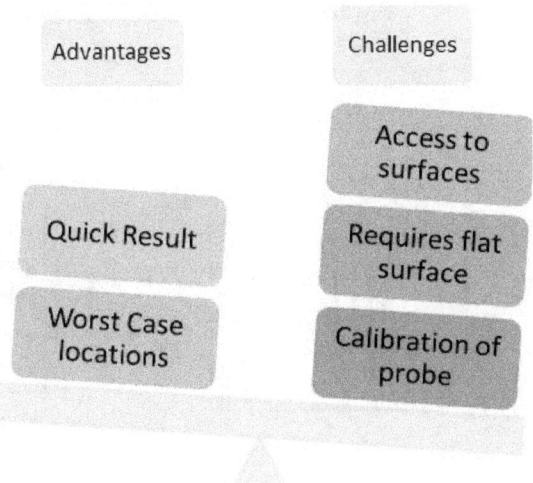

Figure 10: Direct surface sampling (Pros and Cons).

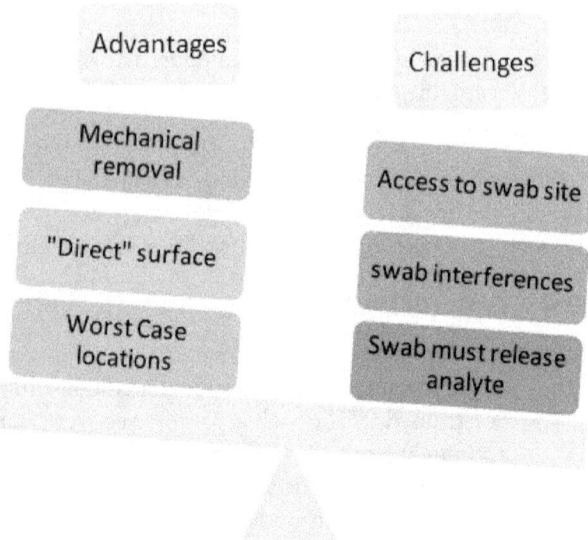

Figure 11: Swab sampling (Pros and Cons)

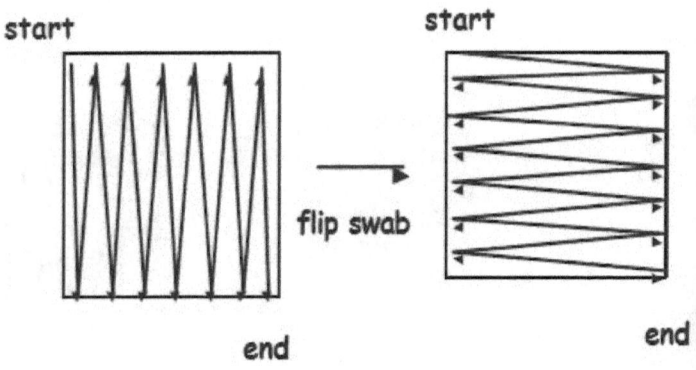

Figure 12: Method of swab sampling.

Rinse Samples

(i) Rinse samples allow sampling of a large surface area. In addition, inaccessible areas of equipment that cannot be routinely disassembled can be evaluated. However, consideration should be given to the solubility of the contaminant.

(ii) A direct measurement of the product residue or contaminant in the relevant solvent should be made when rinse samples are used to validate the cleaning process. Ref: PIC/S PI 006-3.

In rinse sampling a fluid (solvent) is used to rinse and make contact with all surfaces of the item. The sample is then tested quantitatively to remove the target residue.

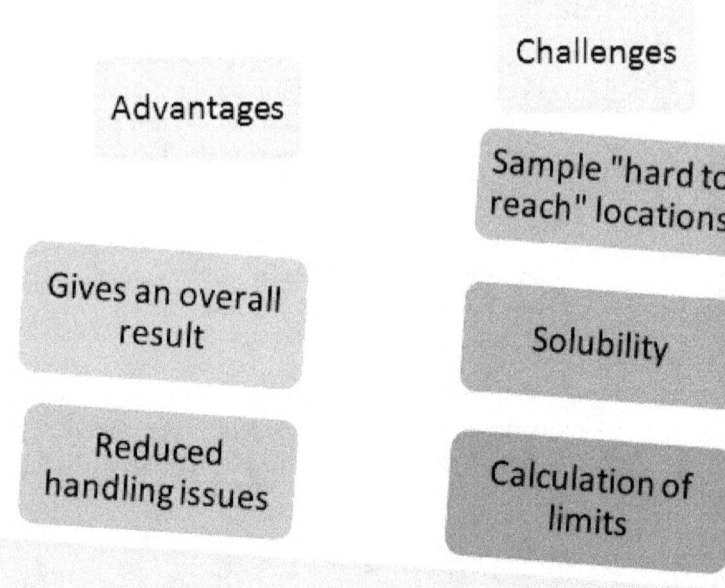

Figure 13: Rinse sampling (Pros and Cons)

<u>Sources of Contaminants</u>

Sampling aims to detect any residue drug content or solvents or other soiling left behind after the cleaning process. The visual inspection is also important in identifying any larger contamination of debris.

Microbial sampling is also done to ensure no microorganisms are present in equipment or in areas of production.

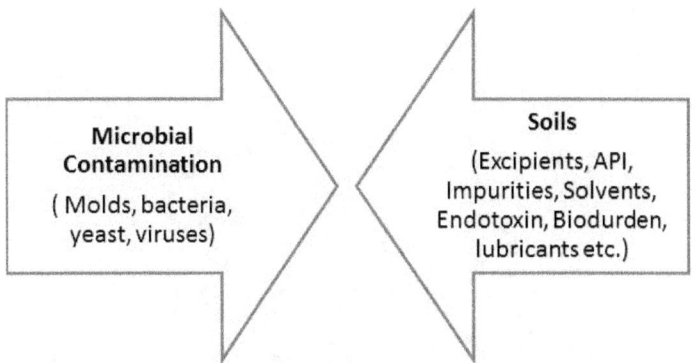

Figure 14: Sources of Contamination

Microbial Sampling

Microbial sampling of utilities such as water-for-injection, purified water, process air etc. is required to ensure no bacteria, moulds, fungi or yeasts are present which risk patient safety.

Colonies can often be determined by visual inspection based on the attributes and appearance test plates/samples. If visual identification is not possible, the colony should be sent for gram stain analysis.

Visual checks involve assessing plates for:

- colour
- shape
- elevation
- size
- texture
- surface
- edge appearance

Examples

Gram Positive Coccus (Staphylococcus Micrococcus Are Examples of GPCs)

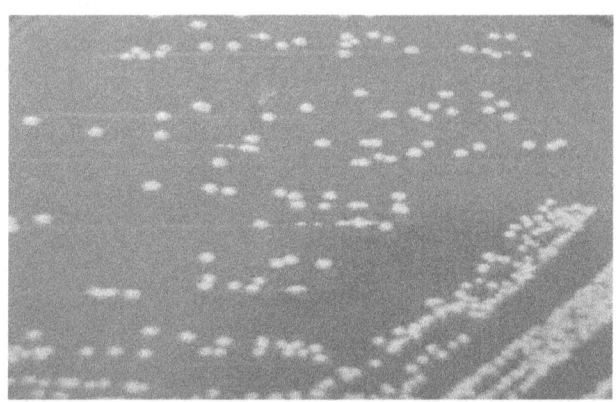

Spherical bacterium, usually slightly less than 1 μ in diameter, coccus belongs to the micrococcaceae family. It is one of the three basic forms of bacteria, the other two being bacillus (ro d-shaped) and spirillum (spiral-shaped).

A pathogenic coccuscan will
almost always be classified as either a staphylococcus (occurr ing in clusters), or a streptococcus (occurring in short orlong chains). Both staphylococci and streptococci are gram- positive and do not form spores.

The staphylococci are responsible for many serious infection s
especially staphylococcus aureus, which is the causativeagen i n boils, abscesses, osteomyelitis, and a large varietyof other i nfections. Staphylococci have received much
attention in recent years because of the ability of most strains to develop a resistance to antibiotics.

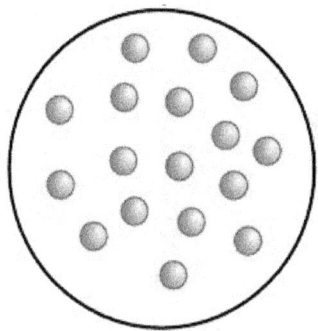

Figure 15: Single cocci.

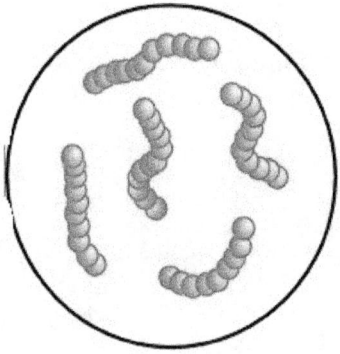

Figure 16: Streptococci.

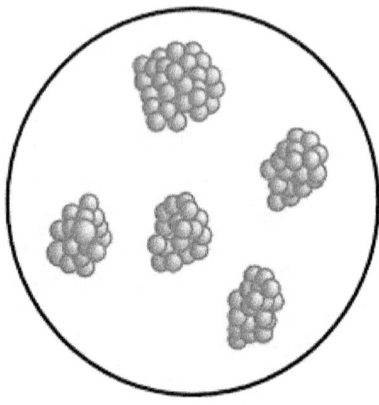

Figure 17: Staphylococci.

Gram-positive rod

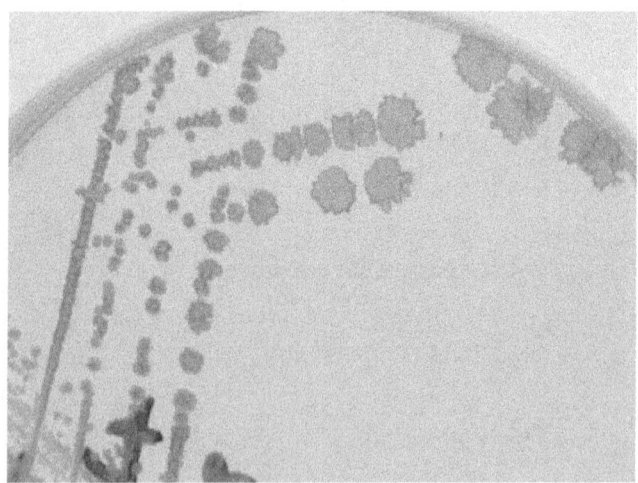

Figure 18: Gram-positive rod

Gram-negative rod (E. Coli)

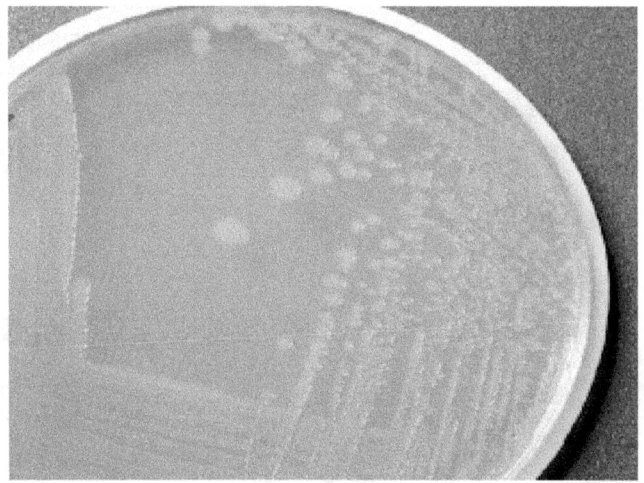

Figure 19: Gram-negative rod (E. Coli)

Gram-negative bacteria are a group of bacteria that do not retain the crystal violet stain used in the gram-staining method of bacterial differentiation.

They are characterised by their cell envelopes, which are composed of a thin peptidoglycan cell wall sandwiched between an inner cytoplasmic cell membrane and a bacterial outer membrane.

Gram-negative bacteria are spread worldwide, in virtually all environments that support life. The gram-negative bacteria include the model organism Escherichia coli, as well as many pathogenic bacteria, such as Pseudomonas aeruginosa and Neisseria gonorrhoeae.

Several classes of antibiotics target gram-negative bacteria specifically, including aminoglycosides and carbapenems.

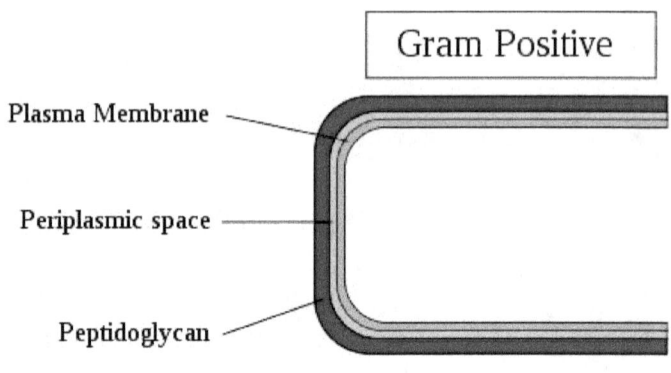

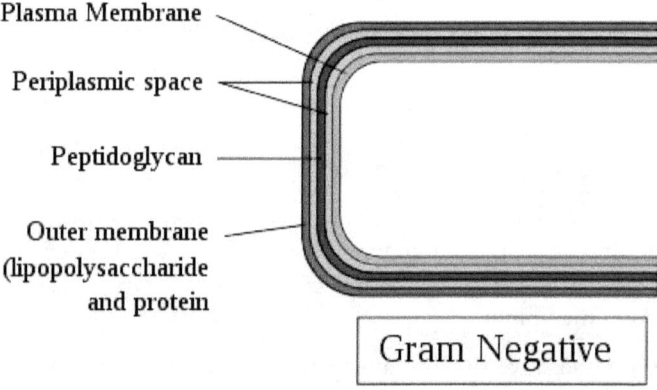

Figure 20: Structure of Gram-Negative Versus Gram Positive.

Objectionable Microorganism Explained

An objectionable microorganism can be defined as a microorganism that can survive or proliferate in a non-sterile drug product or, when appropriate, intermediate manufacturing steps of sterile and non-sterile processes, and adversely affect the appearance, physicochemical attributes or therapeutic effect of that pharmaceutical product. A microorganism that, due to its numbers and pathogenicity, can cause infection, allergic response or toxemia in the patient receiving the product.

Factors that may cause a microorganism to be objectionable:

- Can it cause infection when present in a medicine depending on the route of administration?
- Is it capable of proliferation in a medicine and in excess of the specifications of Total Aerobic Microbial Count (TAMC) or Total combined Yeasts and Moulds Count (TYMC)
- Potential impact of the analytical testing
- Affect finished product, API, excipients and their respective stability
- Produce odours, flavours or undesirable metabolites
- Reduction in the efficacy of the medicine

Atypical Microorganisms Explained

An atypical microorganism can be defined as a microorganism that can survive or proliferate in a sterile finished product and can adversely affect the appearance, physicochemical attributes or therapeutic effect of that pharmaceutical product, and,

A microorganism is one that is not expected to be found in the environment where the product is manufactured and where generally the local flora is dominantly composed of microorganisms of human origin such as gram-positive cocci with staphylococcaceae and corynebacteriaceae.

Frank Pathogen: A microorganism responsible for infection in healthy individuals (i.e. individuals with normal operative and functional host defence mechanisms) that may be acquired from exposure to other infected people or animals, environmental reservoir (exogenous) or the individual's normal (endogenous) microbial flora. [PDA technical report 67].

Microorganism of Concern: A bacterium, yeast or mould that, due to it prominence in product recalls, infection outbreaks, nosocomial infections and the clinical literature, results in a multifactor risk assessment to determine whether the microorganism is objectionable or specified if it is present in a specific non-sterile product or atypical if it is present in a specific sterile product. [PDA TR 67].

Specified Microorganism: Microorganism with limit tests for absence in 1 or 10 g/mL of a finished product, as described in USP<62>/EP 2.6.23 and USP<1111>/EP 5.1.4. [PDA TR 67].

Opportunistic Pathogen: A microorganism responsible for infection in injured, invasively-treated or immune-suppressed individuals that typically does not cause infection in healthy individuals, unlike frank pathogens. [PDA TR 67].

Route of Administration: The way in which a drug product or medicinal device is delivered based on the dosage form and therapeutic use. [PDA TR 67].

Utilities

Introduction

The key utilities involved for cleaning include utilities such as water, compressed gases (air, nitrogen etc.) and the heating and cooling of process equipment. Water quality can impact the effectiveness of pre-rinsing, washing, and final rinsing. Therefore, both the water temperature and quality need to be tightly controlled and monitored. Gases are typically used in order to blowdown or blowout remaining fluids or they are used as a drying step.

The term "clean utilities" in the life science industry refers to utilities that have to fulfil regulatory requirements. The most common utility is water, which can be supplied in different pharmaceutical grades of purity. Purified water (PW or PUW), Highly Purified Water (HPW) and Water-for-Injection (WFI) are the most common. Water quality specifications can be found in the pharmacopeias, e.g. the US Pharmacopeia. Other clean utilities can also include clean compressed air, clean gasses (e.g. nitrogen, argon and oxygen), and clean steam.

Key Definitions

Alert limit: A value reached when the normal operating range of a critical parameter has been exceeded, indicating that corrective measures may need to be taken to prevent the action limit being reached.

At-Rest: A condition where the installation is complete with equipment installed and operating in a manner agreed upon by the customer and supplier, but with no personnel present.

Cleanroom: an area (or room or zone) with defined environmental control of particulate and microbial contamination, constructed and used in such a way as to reduce the introduction, generation and retention of contaminants within the area.

Containment: A process or device to contain product, dust or contaminants in one zone,
preventing it from escaping to another zone.

Contamination: The undesired introduction of impurities of a chemical or microbial nature, or of foreign matter, into or onto a starting material or intermediate, during production, sampling, packaging or repackaging, storage or transport.

Point extraction: Air extraction to remove dust with the extraction point located as close as possible to the source of the dust.

Pressure cascade: A process whereby air flows from one area, which is maintained at a higher pressure, to another area at a lower pressure.

Relative humidity: The ratio of the actual water vapour pressure of the air to the saturated water vapour pressure of the air at the same temperature expressed as a percentage. More simply put, it is the ratio of the mass of moisture in the air, relative to the mass at 100% moisture saturation, at a given temperature.

Turbulent flow: Turbulent flow, or non-unidirectional airflow, is air distribution that is introduced into the controlled space and then mixes with room air by means of induction.

The process of identifying critical utilities can be done with the application of direct impact, indirect impact and no impact definitions. Risk assessments, CQAs and CPPs should also help identify critical utilities. When critical utilities are required as part of manufacturing and processing, the following points should be examined during the requirements and design stage:

- Materials of construction
- Internal surface finishes
- System sizing
- Flow rates, dead legs, drainage etc.

<u>Compressed Air</u>

Compressed air is used for valve actuation, instrument air and process air to name but a few applications. Only the point-of-use filtration and the gas quality instrumentation should be classified as level 1. When flow or pressure is a CPP, the measurement/monitoring should be performed by the system into which the gas is flowing. Additionally, the CQAs and CPPs should be routinely monitored through the calibrated monitoring system. For compressed air, the potential CPPs are listed below. For the physical system being evaluated, the use and the application of the compressed air will determine which (if not all) CPPs are needed to ensure the system produces product of the desired quality.

- ➢ Hydrocarbons
- ➢ Moisture
- ➢ Particulates
- ➢ Temperature

It is important that each point of use has appropriate sterile filters in place. If the filter is not placed directly at the point of use, control and counter measures should be implemented to address any risk of contamination downstream of the filter. Compressed air for bio-pharmaceutical use must be generated using oil free compressors with appropriate temperature controls in place.

Attribute	Clean Compressed Air (impacts product quality)	Sterile Compressed Air (impacts sterile product quality)
Oil content	*Not great than 0.1mg/m^3	
Microbiological requirement	Meets requirements of the environmental zone served (e.g. Grade B,C etc.)	Sterile
Filtration requirement	Minimum 0.45µm membrane filter	0.2µm membrane filter

*ISO 8573-1 Class 2

Water Systems

Water supply and the associated water Systems in biotechnology and pharmaceuticals are a vital component of the manufacturing process. They are used to clean equipment and vessels, to cool or heat processing pipes and systems, and in many circumstances certain grades of water are a component of the finished product (e.g. water-for-injection). Various grades of water service have a particular purpose.

Some common types include:

- Potable water
- Soft water
- Purified water
- Water-for-injection

Water used in process and in cleaning should be pure and free from microbial and chemical impurities. As the water gets easily contaminated by environmental conditions, diligence in the design is essential. Typically water systems are supplied on a continuous loop with recirculation.

CPPs typical for a water system include:

- Pressure
- pH
- Conductivity
- Level
- TOC
- Flow
- Temperature
- Resistivity

Water for Injection:

WFI is sterile and pyrogen-free water containing no less than 10 CFU/100ml (Colony Forming Units) with a sample size of between 100 and 300 ml and an endotoxin level < 0.25 EU/ml.

The use of WFI is two-fold. Firstly, it can be used for critical processing steps such as washing and rinsing .It can also be used in injectable products. WFI is a key raw material for sterile intravenous and intradermal products. WFI is produced by a Multi-Column Distillation Plant (MCDP), and must meet the microbial requirements of regulated bodies.

Clean-in-Place (CIP) / Sterilise-in-Place (SIP) System

The cleaning of equipment, vessels and process piping is a critical activity. Any residue from a previous production batch needs to be removed in order to avoid cross-contamination. CIP and SIP skids are often utilised to allow efficient switchover between batches and/or products.

Where possible, manual cleaning should be avoided unless essential due to the design of a system or particular location or configuration.

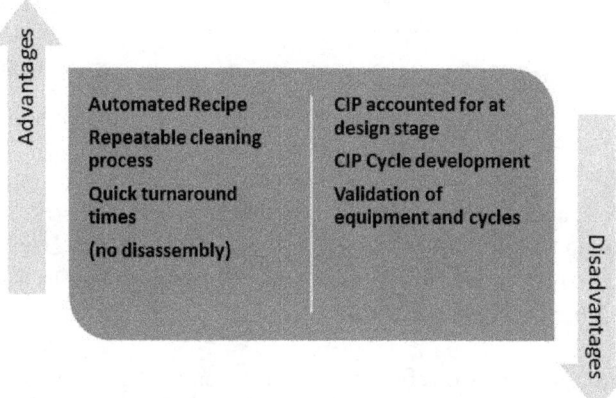

Figure 21: Advantages and disadvantages of CIP.

Clean Steam

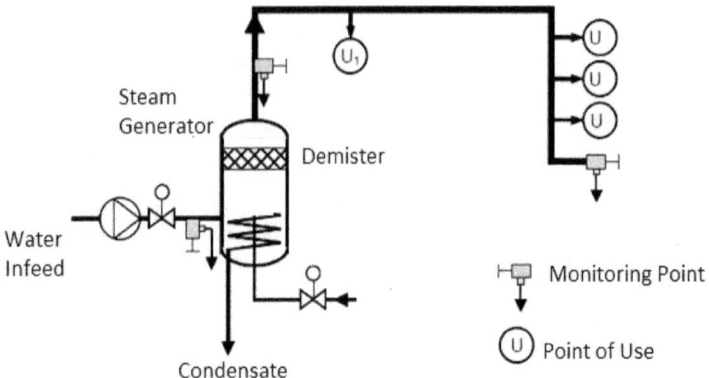

Figure 22: Simple Clean Steam Generation Piping and Instrumentation.

Pure steam is used in pharma and biotech for sterile application, for autoclave sterilisation etc. Distribution piping of clean steam is a critical aspect. Improper sizing of pipes may impact the production process and lead to a loss of time during sterilisation.

Clean steam, also referred to as "pure steam", and gases used in manufacturing operations must be of a quality suitable for their intended purpose. The intended use of clean steam and gases must be understood in order to determine any risks to the patient or product. For example, gases that end up being part of the product must fulfil the regulatory requirements. Preventative maintenance and ongoing monitoring must be implemented for clean steam systems.

- ➢ Routine inspection and maintenance.
- ➢ Frequency of filter change
- ➢ Frequency of the sterilisation for the gas distribution system, if applicable

> Frequency for integrity testing of the sterile filter

Water systems for purified water, de-ionised water and Water-for-Injection (WFI) must provide a consistent and reproducible output. Where there is moisture, there is always a risk of microbial contamination. Therefore, the design of water systems should mitigate against such risks. Good engineering practices such as using circulation loops, no dead legs and polished-surface finishes all work to provide an effective and safe system. The design should also take into account ease of sampling at the point of use. The removal of endotoxins is a requirement for WFI.

Ongoing sampling monitoring the quality of water is particularly important where water systems are concerned. Procedures should be in place to ensure that effective monitoring and testing is maintained. Action limits and acceptance criteria should be clearly documented in approved SOPs or the equivalent. Failure to meet limits or acceptance criteria should initiate an investigation. The potential CPPs are listed below for clean steam systems:

> Conductivity
> Flow
> Level
> Pressure
> Resistivity
> Temperature

Design Considerations:

The purpose of a User Requirement Specification (URS) is to define the requirements for the operation and control of the clean steam system:

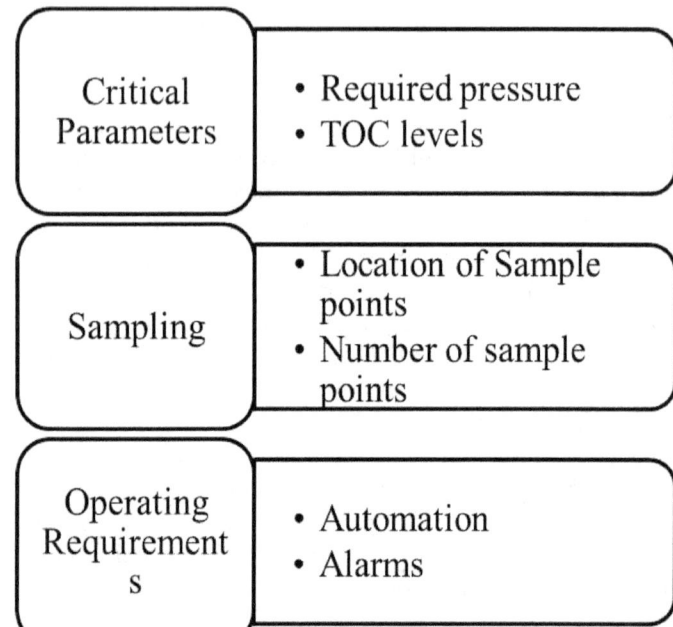

OQ Testing

Operational qualification or OQ is a formal validation activity, and as such should be completed per an approved protocol. The purpose of OQ testing is to confirm the operational and functionality of the clean steam system. This should demonstrate that all critical aspects of a URS are fulfilled.

Key verifications include:

➢ Testing of temperatures and operating pressures
➢ Capacity testing (under load)
➢ Steam trap operation
➢ Verification of automated functions and alarms
➢ Check of automation systems, including PCS

> Correct function of valves and sampling points

PQ Testing

Due to the high operating temperatures and the associated lethality, clean steam systems are resistant to microbiological contamination.

Issues that arise can normally be attributed to equipment failures with the steam generator or contaminated water being supplied to the system. Bacterial endotoxin testing is used to monitor clean steam systems for both PQ purposes and throughout the life cycle of the equipment operation. Steam is condensed, sampled and tested. The condensate should meet WFI specifications with the exception of viable total aerobic count.

Clean steam PQs are commonly completed using a three-phase approach to testing. The first phase ensures the system consistently operates within the required ranges and the steam provided meets the acceptance criteria. Typically, phase one bacterial endotoxin testing and physio-chemical testing is completed over a two week period. For phase two, the same frequency and type of testing may be applied for an additional two weeks. After phase two testing, the system may be available for general use if allowed for within internal company procedures. Phase two testing at PQ should also provide a report with all results documented and reviewed.

Phase three of PQ is intended to demonstrate the effective and consistent operation of the system over a longer term (approx. 12 months). Sampling is typically performed weekly.

Further Reading on Clean Steam

- PIC/S PI009-3 – Pharmaceutical Inspection Co-Operation Scheme - Inspection of Utilities

- EN 285 – European Standard - Sterilisation, Steam Sterilisers, Large Sterilisers

- USP <1231> – United States Pharmacopeia - <1231> "Water for Pharmaceutical Purposes"

- USP– United States Pharmacopeia - Monograph "Pure Steam"

EN 285 – European Standard - Sterilisation, Steam Sterilisers, Large Sterilisers

Case Study 1 – Example of Production Line Switching and Comprehensive Cleaning Procedure/ Requirements

General Safety

1. Switch off the machine prior to commencement of any cleaning operation.

2. LOTO and the authorised worker tag is to be applied by the authorised worker with reference to the applicable LOTO procedure

3. Before starting machine cycles, ensure that no operators are working on the machine and that all operators have been informed and acknowledge the machine will be in cycle.

4. Complete GMP permits if required by local and site procedures.

5. Place appropriate signage in-situ.

6. Ensure all guards are replaced and checked prior to start-up on the line.

7. Do not allow water to enter electrical equipment when cleaning.

8. Wear nitrile gloves when handling chemical detergents.

150 Tonne V-Blender - Cleaning Procedure

General cleaning on a daily basis when in production:

1. Wipe down or vacuum the outside surface of the blender in the area around the access covers and around the discharge valve so that no obvious dust accumulation remains.

2. Comprehensive cleaning.

A comprehensive clean is carried out:

   i) After 40 calendar days where consecutive operations of the same product take place.

   ii) Between batches of different product.

   iii) As soon as possible after use if the equipment is to be idle for more than one week.

   iv) If equipment has been comprehensively cleaned and is idle for more than seven days, then a re-rinsewith purified water is sufficient

V-Blender - Manufacturing Areas

Turn off the mains isolator on the blender control panel.

Note: All equipment must be isolated from the mains before commencing a comprehensive clean.

All electrical panels, plugs and sockets must be covered and sealed off. Spraying of water near electrical equipment must be avoided.

Clean the outer and inner surfaces of the blender and all detachable components as follows:

i) Wash with a solution made up with the current approved detergent as per cleaning policy and hot water.

ii) A bucket, cloth and mop can be used to apply the detergent solution.

iii) Ensure that all product residues are removed.

vi) Rinse the washed equipment with hot tap water twice using a bucket, cloth and brush. Make sure that there is no visible trace of product on the equipment.

vi) Rinse the agitator bar and internal blender surfaces with purified water making sure that there are no visible traces of detergent on the equipment and allow to air dry.

viii) Wipe off excess water from the outer blender surfaces.

Re-assemble the equipment only if it is to be used within seven days of cleaning.

Log the cleaning in the appropriate logbook.

A post comprehensive clean checklist must also be completed by both the operator and the team leader.

Case Study 2 – Coating Pan

Coating pans are used to coat tablets with thin films of material to enhance a particular property of a tablet. Some coatings are used to slow the dissolution rate, while others may simply act to allow easier swallowing for the patient such as carnauba wax which is a safe, non-toxic and inert ingredient.

The swab sites should be adequate in number and location to provide representative testing, taking into account both easy-to-clean and hard-to-clean locations.

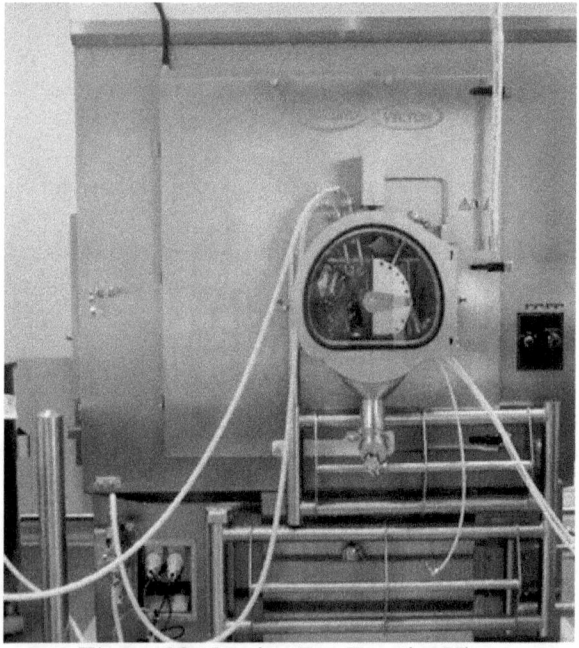

Figure 23: Coating Pan Exterior View

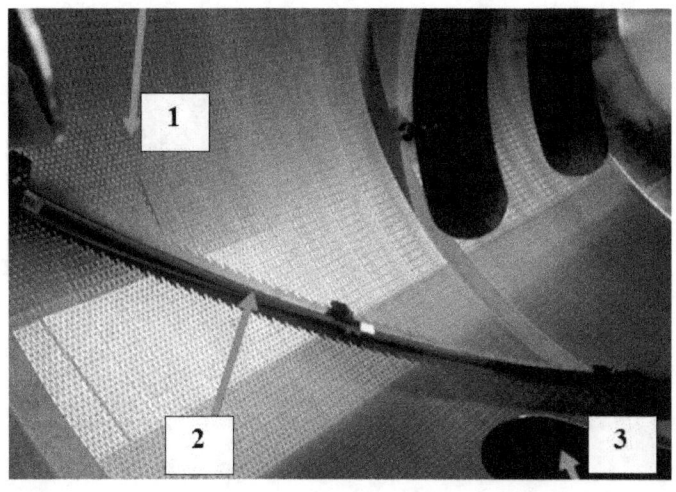
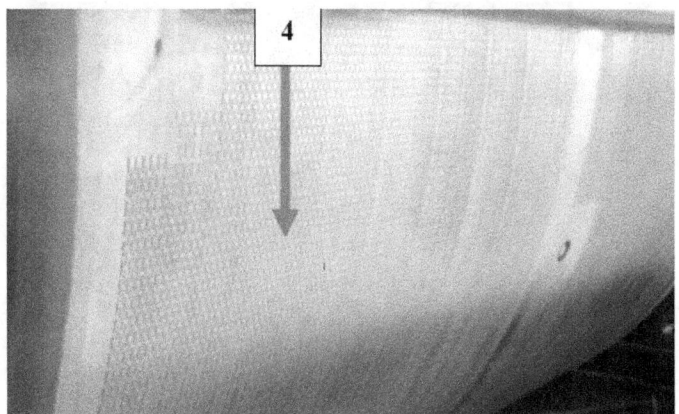

Swab Site 1 – Hard-to-clean, interior of perforated pan

Swab Site 2 – Hard-to-clean, inner side of wide baffles
Swab Site 3 – Hard-to-clean, surface of narrow baffles
Swab Site 4 – Hard-to-clean, outer surface of pan

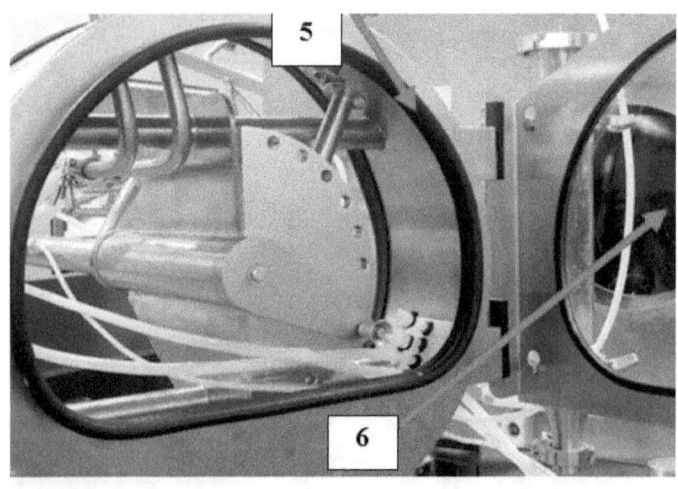

Swab Site 5 – Easy-to-clean, rubber seal on door
Swab Site 6 – Easy-to-clean, door of pan
Swab Site 7 – Hard-to-clean, inside of solution IN line
Swab Site 8 – Hard-to-clean, inside of solution OUT line

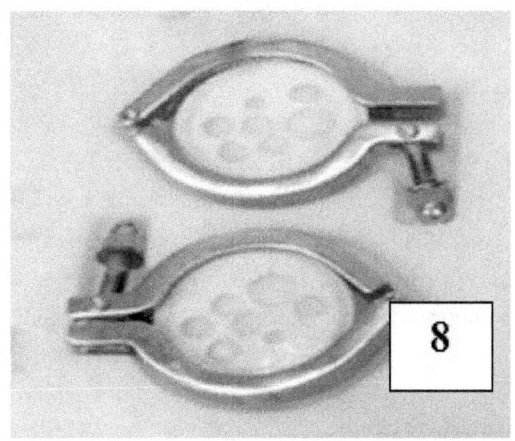

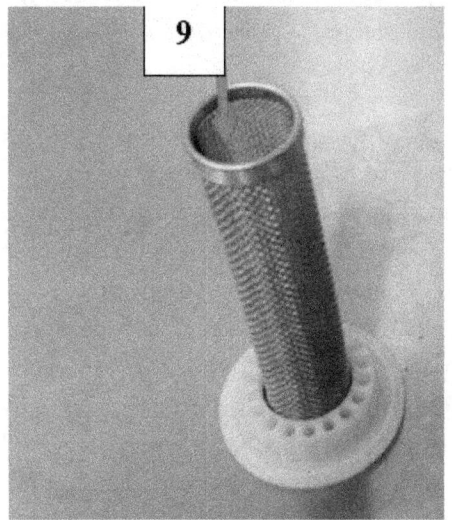

Swab Site 8 – Easy-to-Clean, rubber seal rim
Swab Site 9 – Hard to Clean, inside the filter housing

Note: Gun assemblies and ancilary equipment may also be included in a validation based on risk.

Inspection of Cleaning Processes

Introduction

As far back as 1963 GMP Regulations (Part 133.4), the FDA required equipment to be clean. The regulations stated that equipment "shall be maintained in a clean and orderly manner." Nowadays, the FDA is mostly concerned with the cross-contamination of drug products with potent steroids or hormones.

A historical example of cross-contamination due to inadequate procedures was the 1988 recall of a finished drug product, Cholestyramine resin USP.

"The bulk pharmaceutical chemical used to produce the product had become contaminated with low levels of intermediates and degradants from the production of agricultural pesticides. The cross-contamination in that case is believed to have been due to the re-use of recovered solvents. The recovered solvents had been contaminated because of a lack of control over the reuse of solvent drums. Drums that had been used to store recovered solvents from a pesticide production process were later used to store recovered solvents used for the resin manufacturing process. The firm did not have adequate controls over these solvent drums, did not do adequate testing of drummed solvents, and did not have validated cleaning procedures for the drums.

Some shipments of this pesticide contaminated bulk pharmaceutical were supplied to a second facility at a different location for finishing. This resulted in the contamination of the bags used in that facility's fluid bed dryers with pesticide contamination. This in turn led to cross contamination of lots produced at that site, a site where no pesticides were normally produced.

FDA instituted an import alert in 1992 on a foreign bulk pharmaceutical manufacturer which manufactured potent steroid products as well as non-steroidal products using common equipment. This firm was a multi-use bulk pharmaceutical facility. FDA considered the potential for cross-contamination to be significant and to pose a serious health risk to the public. The firm had only recently started a cleaning validation programme at the time of the inspection and it was considered inadequate by FDA. One of the reasons it was considered inadequate was that the firm was only looking for evidence of the absence of the previous compound. The firm had evidence, from TLC tests on the rinse water, of the presence of residues of reaction by-products and degradants from the previous process."

Ref:
https://www.fda.gov/iceci/inspections/inspectionguides/

General Requirements

Written procedures (aka SOPs) are the starting point for application of any cleaning programme within industry. The procedures should give adequate detail on the cleaning processes used for various pieces of equipment.

If different cleaning processes exist for changing between different products then these must be described. Similarly, if there is one process for removing water soluble residues and another process for non-water soluble residues, the written procedure should address both scenarios.

> *"FDA expects firms to have written general procedures on how cleaning processes will be validated.*

> *FDA expects the general validation procedures to address who is responsible for performing and approving the validation study, the acceptance criteria, and when revalidation will be required.*

> *FDA expects firms to prepare specific written validation protocols in advance for the studies to be performed on each manufacturing system or piece of equipment which should address such issues as sampling procedures, and analytical methods to be used including the sensitivity of those methods.*

> *FDA expects firms to conduct the validation studies in accordance with the protocols and to document the results of studies.*

> *FDA expects a final validation report which is approved by management and which states whether or not the cleaning process is valid. The data should support a conclusion that residues have been reduced to an "acceptable level."*

Evaluating Cleaning Validation

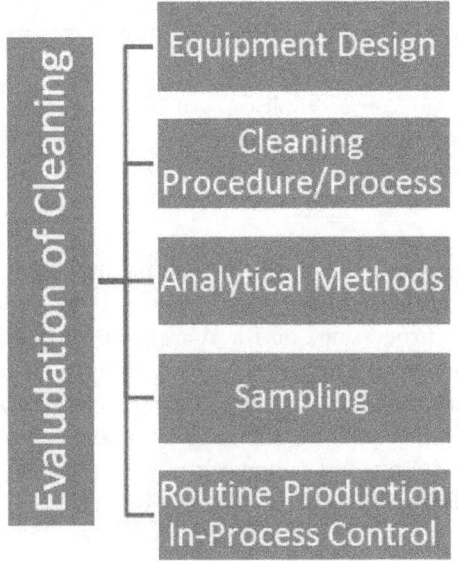

Figure 24: Key Elements of How Cleaning Validation Is Evaluated.

Equipment Design

It is important to examine the design of equipment, particularly in those large systems that may employ semi-automatic or fully automatic clean-in-place (CIP) systems since they represent significant concern. For example, sanitary type piping without ball valves should be used.

Auditors and inspectors also look at the level of training and experience in cleaning these systems. It's equally important to check the written and validated cleaning process to determine if these systems have been properly identified and validated. Always check for the presence of an often critical element in the documentation of the cleaning processes; identifying and controlling the length of time between the end of processing and each cleaning step. This is especially important for topicals, suspensions, and bulk drug operations. In such operations, the drying of residues will directly affect the efficiency of a cleaning process.

After, cleaning equipment may be subjected to sterilisation or sanitisation procedures where such equipment is used for sterile processing, or for non-sterile processing where the products may support microbial growth. Therefore, control of the bioburden through adequate cleaning and storage of equipment is important to ensure that subsequent sterilisation and sanitization procedures achieve the necessary assurance of sterility.

<u>Cleaning Process, Procedure and Documentation</u>

Examine the detail and specificity of the procedure for the (cleaning) process being validated, and the amount of documentation required. We have seen general SOPs, while others use a batch record or log sheet system that requires some type of specific documentation for performing each step. Depending upon the complexity of the system and cleaning process and the ability and training of operators, the amount of documentation necessary for executing various cleaning steps or procedures will vary.

When more complex cleaning procedures are required, it is important to document the critical cleaning steps (for example certain bulk drug synthesis processes). In this regard, specific documentation on the equipment itself which includes information about *who* cleaned it and *when* is valuable. However, for relatively simple cleaning operations, the mere documentation that the overall cleaning process was performed might be sufficient.

Analytical Methods

The manufacturer should determine the specificity and sensitivity of the analytical method used to detect residuals or contaminants. If levels of contamination or residual are not detected, it does not mean that there is no residual contaminant present after cleaning. It only means that levels of contaminant greater than the sensitivity or detection limit of the analytical method are not present in the sample. The firm should challenge the analytical method in combination with the sampling method(s) used to show that contaminants can be recovered from the equipment surface and at what level, i.e. 50% recovery, 90%, etc. This is necessary before any conclusions can be made based on the sample results. A negative test may also be the result of poor sampling technique.

Sampling

There are two general types of sampling that have been found acceptable. The most desirable is the direct method of sampling the surface of the equipment. Another method is the use of rinse solutions.

Advantages of direct sampling are that areas hardest to clean and which are reasonably accessible can be evaluated, leading to establishing a level of contamination or residue per given surface area. Additionally, residues that are "dried out" or are insoluble can be sampled by physical removal.

Two advantages of using rinse samples are that a larger surface area may be sampled and inaccessible systems or ones that cannot be routinely disassembled can be sampled and evaluated.

A disadvantage of rinse samples is that the residue or contaminant may not be soluble or may be physically occluded in the equipment. An analogy that can be used is the "dirty pot." In the evaluation of cleaning of a dirty pot, particularly with dried out residue, one does not look at the rinse water to see that it is clean; one looks at the pot.

Routine Production In-Process Control

Monitoring - Indirect testing, such as conductivity testing, may be of some value for routine monitoring once a cleaning process has been validated. This would be particularly true for the bulk drug substance manufacturer where reactors and centrifuges and piping between such large equipment can be sampled only using rinse solution samples. Any indirect test method must have been shown to correlate with the condition of the equipment. During validation, the firm should document that testing the uncleaned equipment gives a non- acceptable result for the indirect test.

Establishment of Limits

The FDA or other regulatory bodies do not intend to set acceptance specifications or methods for determining whether a cleaning process is validated. However, a company's rationale for the residue limits established should be logical based on the manufacturer's knowledge of the materials involved and be practical, achievable, and verifiable. It is important to define the sensitivity of the analytical methods in order to set reasonable limits. Some limits that have been mentioned by industry representatives in the literature or in presentations include analytical detection levels such as 10 PPM, biological activity levels such as 1/1000 of the normal therapeutic dose and organoleptic levels such as no visible residue.

Useful References

- FDA Process Validation: General Principles and Practices (Appendix A)

- 21 CFR 820.75 Process Validation

- EN ISO 13485:2012 Medical Devices – Quality Management Systems - Requirements for Regulatory purposes (ISO 13485:2003)

- http://www.picscheme.org/publication.php?id=4

- ASME BPE-2000 (Welding) Standard

- FDA – Food and Drug Administration - Guide to Inspections of Validation of Cleaning Processes

- EU GMP – European Commission – EudraLex Volume 4: EU Guidelines to Good Manufacturing Practice, Medicinal Products for Human and Veterinary Use, and Annex 15 (section 10 "Cleaning Validation")

- ICH Q7 – International Council on Harmonisation - Good Manufacturing Practice
- Guide for Active Pharmaceutical Ingredients (section 12.7 "Cleaning Validation")

- ICH Q9 – International Council on Harmonisation - Quality Risk Management

- PIC/S PI 006-3 – Pharmaceutical Inspection Co-operation Scheme -Recommendations on Validation Master Plan, Installation and Operational Qualification, Non-Sterile Process Validation, Cleaning Validation (section 7 "Cleaning Validation")

- WHO TRS 937 – World Health Organisation - Specifications for Pharmaceutical Preparations; Annex 4: Supplementary Guidelines on Good Manufacturing Practices: Validation; Appendix 3: Cleaning Validation

- ICH Q3C(R5) – International Council on Harmonisation – Impurities: Guideline for Residual Solvents

- WHO TRS 986 annex 2 – World Health Organisation – WHO Good Manufacturing Practices for Pharmaceutical Products: Main Principles

- WHO TRS 937 – World Health Organisation - Specifications for Pharmaceutical Preparations; Annex 4: Supplementary Guidelines on Good Manufacturing Practices: Validation; Appendix 3: Cleaning Validation

- Health Canada Guide-0028– Health Canada - Cleaning Validation Guidelines

- PDA TR29– Parenteral Drug Association – TR29 Points to Consider for Cleaning Validation

Appendix I

Precision Cleaning (Medical Devices)

Load	Clean	Rinse	Clean	Rinse	Rinse	Dryer
	DI Water	DI Water	Nitric Acid with DI Water	DI Water	DI Water	
						Hot Air Blower
	Detergent	Heat, Agitation & Ultrasonics	Heating and Ultrasonics	Ultrasonics and heating	Ultrasonics and heating	

Precision cleaning equipment such as ultrasonic aqueous clean1 lines or solvent based degreasers have one fundamental function in common. This is to remove or reduce particulate, grease and dirt from parts or components. Typically, aqueous or solvent based systems can be used to clean the likes of metallic hip and knee implants, metallic fixation devices, and surgical tools.

Precision cleaning can be divided into two sub-categories, intermediate cleaning processes and final cleaning processes. As the name may suggest, intermediate cleaning processes are less critical than final cleaning processes and often a reduction in soiling levels (aka organic residuals) is the preferred acceptance criteria. Whereas with final cleaning processes, the level of "cleanliness" is greater and therefore specific acceptance criteria are applied, based on the application and design of the medical device.

Glossary

A

Accelerated Ageing

When the deterioration of a device or product component from natural ageing is accelerated and simulated in the laboratory.

Accuracy

Accuracy or trueness. An expression of the closeness of agreement between the value that is accepted, either as a conventional true value or an accepted reference value and the value obtained. A system with low bias implies good accuracy and vice versa.

Adverse Event

A situation or condition that occurs when a data point, result, or process etc. is outside the expected or predetermined limits or ranges.

Affinity Chromatography

A chromatography separation method based on a chemical interaction specific to the target species. Types of affinity methods are: bio sorption, site recognition (e.g. monoclonal antibody, protein A); hydrophobic interaction and contacts between non-polar regions in aqueous solutions.

Air Exchange Rate per Hour (ACPH)

The rate of air exchange expressed as the number of air changes per hour and calculated by dividing the volume of air delivered in the unit of time by the volume of space.

Active Pharmaceutical Ingredient

Any substance or mixture of substances intended to be used in manufacturing a drug (medicinal) product and that, when used in the production of a drug, becomes an active ingredient of the drug product. Such substances are intended to furnish pharmacological activity or other direct effect in the diagnosis, cure, mitigation, treatment, or prevention of disease, or to affect the structure and function of the body. (ICH Q7A, Annex 18, Part II).

ANSI

American National Standards Institute

Antimicrobial Resistance

Antimicrobial resistance corresponding to the emergence and spread of microbes that are resistant to cheap and effective first-choice, or "first-line" antimicrobial drugs.

Application

A term most often used in relation to software validation and computerised systems. It is any software installed on a defined platform providing specific functionality.

Approve

"Approve" the device after reviewing a premarket approval (PMA) application that has been submitted to FDA.

AVL (Approved Vendor List)

A list of all the vendors or suppliers approved by a company as sources from which to purchase materials.

Artwork

Electronic files or print outs containing the representation of a packaging item, graphical elements and regulatory text. Approved artworks are used by suppliers for printing.

Aseptic (Conditions)

Conditions in which the working environment under which the potential for microbial and/or viral contamination are minimised.

ASTM

American Society for Testing and Materials.

ATEX

An acronym derived from the French-titled 'Atmosphères Explosibles' 94/9/EC directive outlining what equipment and work environment is allowed in an environment with an explosive atmosphere. This European directive amends and adds safety requirements for hazardous areas in the relevant national legislation in the member states of the European Union, bringing in a common standard. Where equipment is to be used in potentially explosive atmospheres containing gas or combustible dust, it must comply with the ATEX directive.

Audit Trail

The audit trail is a control mechanism of a system that allows all data entered or modified to be traced back to the original data. A reliable and secure audit trail is particularly important in conjunction with the creation, change or deletion of GMP-relevant electronic records.

Acceptable Quality Level (AQL)

The AQL of a sampling plan is the process performance level routinely accepted by the sampling plan.

B

Basis of Design

A design document that demonstrates a thorough understanding of the project and its intended output. Typically, it contains of preliminary drawings and system descriptions etc. Together with the URS and the detailed design, it provides overall evidence that the design addresses the requirements of the equipment, system or facility.

Biocompatibility

A measure of how a biomaterial interacts in the body with the surrounding cells, tissues and other factors.

Bioburden

The level and type of micro-organisms that can be present in raw materials, API starting materials, intermediates or APIs. Bioburden should not be considered contamination unless the levels have been exceeded or defined objectionable organisms have been detected.

Biological Indicators

Test system containing viable microorganisms providing a defined resistance to a specified sterilisation process, e.g. vaporised hydrogen peroxide.

Biomaterial

Any matter, surface, or construct that interacts with biological systems. Biomaterials can be derived from nature or synthetic (manufactured). The active substance of a biosimilar medicine is comparable to a biological reference medicine. Biosimilar and biological reference medicines are used at the same dose to treat the same disease. The name, appearance and packaging of a biosimilar medicine differs to that of a biological reference medicine.

Bracketing

A bracketing (aka family or matrix) approach can be used where similar products are produced using the same equipment and processes. A particular product size or product configuration may be selected to represent the worst-case product. Therefore, by qualifying the worst case, all of the other products within the family are considered validated.

Body Orifice

Any natural opening in the body, as well as the external surface of the eyeball, or any permanent artificial opening, such as a stoma or permanent tracheotomy.

Borderline Classifications

In certain circumstances, it may not be clear if a product falls under the medical device legislation or whether to classify a device as a medicine, cosmetic, biocide and so on. The decision will largely depend on the particular intended use of the product, as assigned by the manufacturer, and on the demonstrated mode of action. The manufacturer's claims must be substantiated by relevant data.

Bulk Product

Any pharmaceutical form (liquid, powder, suspension) that is to be filled into either another container or its final container at the next process step; or is already filled into its final container to be labelled and packaged at the next process step.

BOM

Bill of Materials.

BSI

British Standards Institute.

C

CAD, Computer Aided Drawing

A system used to create physical designs, usually three-dimensional. Some examples of CAD software are SolidWorks, Pro/ENGINEER and AutoCAD.

Calibration

A requirement that demonstrates that a particular instrument or device produces results within specified limits by comparison with those produced by a reference or traceable standard over an appropriate range of measurements.

Campaign (Process)

A production strategy where consecutive batches of an API, a finished product, or intermediates are processed before the production line/system is cleaned.

Capability (Process Capability)

Process capability is a measure of how capable the process is of producing product meeting specified requirements. It is a measure of the actual variation in that product characteristic compared to the product specifications. Indices are used to represent the process capability such as Pp, Cp and Ppk, Cpk, depending on how the data is collected e.g. multiple batches over time.

CAPA

A Corrective and Preventive Action. A systematic approach that includes actions needed to correct, prevent recurrence and eliminate the cause of potential non-conforming product and other quality problems (preventive action). (21CFR 820.100).

Change Control

A formal system by which qualified representatives of appropriate disciplines review proposed or actual changes that may impact the validated status.

Change Notification (Agreement)

A signed declaration that states that the supplier agrees to notify the customer of changes in its product or process in order to allow the customer determine whether the changes can affect the quality of finished goods or a quality system.

Change Management

An overarching approach to change control that is used during the preliminary planning and design stage of a project.

Cleaning

The process of removing potential contaminants from process equipment and maintaining the condition of the equipment so that it can be safely used for subsequent product manufacture.

Cleaning Validation

Documented evidence that provides a high degree of assurance that a specific cleaning process will consistently produce a result meeting predetermined requirements for cleanliness.

Cleaning Verification

Confirmation by examination and provision of objective evidence that specific requirements have been fulfilled.

Cocurrent (Flow)

This is when the fluids are applied in the same direction. Cocurrent flow is less effective as less heat can be transferred; therefore it is less commonly used.

Code of Federal Regulations (CFR)

Regulations issued by U.S. government agencies. The individual titles making up the regulations are numbered the same way as the federal laws on the same topic.

Competent Authority

A competent authority is the legally designated authority mandated to monitor compliance with directives and legal requirements within the industry. The competent authority has the power to grant and revoke licenses.

Compendial Organisations

Organisations certifying material standards that meet compendial requirements and acceptance criteria. (E.g. USP).

Commissioning

An engineering activity that includes all aspects of overseeing a system, piece of equipment or process through to successful installation until ready for use. Commissioning involves both requirements of installation qualification (IQ) and operational qualification (OQ).

Computer System

A group of hardware components and associated software designed and assembled to perform a specific function or group of functions.
[EU GMP Guide, Part II, ICH Q7].

Computerised System

A system including the input of data, electronic processing and the output of information to be used either for reporting or automatic control. [EU GMP Guide, Glossary].

Computer System Validation

A process that confirms by examination and provision of objective evidence that the computer system conforms to user needs and intended uses. System validation is a process for achieving and maintaining compliance with GxP regulations and fitness for intended use by adoption of life cycle activities, deliverables, and controls.

Concurrent Validation

Concurrent validation occurs when activities are executed at the same time as one another or concurrent to a product launch.

Confidence Level

Confidence level is expressed as a percentage and represents the probability that the conclusion of the test is correct. A 95% confidence level means that you can be 95% certain that the conclusion is correct.

Conflict of Interest

A conflict of interest is a situation in which a public official's decisions are influenced by the official's personal interests.

Continual Improvement, CI

Ongoing activities to evaluate and positively change products, processes, and the quality system to increase effectiveness

Consent Decree

A consent decree is a binding order issued by a judge that stipulates the voluntary agreement by the participants in a case of litigation. Decrees are sometimes issued after one party voluntarily agrees to cease a particular action without admitting to any illegality of the action to date.

Colony Forming Unit

One or more microorganisms that produce a visible, discrete growth on an agar-based microbiological medium.

Controlled Substances

Products that are categorized due to their potential for abuse, medical use and requirement for medical supervision.

Controlled classified areas

An environment supplied with HEPA-filtered air where materials, equipment, and personnel are regulated to control viable and non-viable particulates to an acceptably low level. Such areas are classified according to the maximum level of airborne particulate allowed.

CNC (Controlled Not Classified)

While these are not ISO recognised room classes, they are generally used to describe non-GMP areas with a level of control in effect.

Clear (FDA)

The attainment of FDA 'clearance' for the device after reviewing a premarket notification, otherwise known as a 510(k) (named after a section in the Food, Drug, and Cosmetic Act) that has been filed with FDA.

Clean Room

An area (or room or zone) with defined environmental control of particulate and microbial contamination, constructed and used in such a way as to reduce the introduction, generation and retention of contaminants within the area.

Containment

A process or device to contain product, dust or contaminants in one zone, preventing it from escaping to another zone.

Contamination

The undesired introduction of impurities of a chemical or microbial nature, or of foreign matter, into or onto a starting material or intermediate, during production, sampling, packaging or repackaging, storage or transport.

Continued Process Verification

Once the initial validation is completed, it is important that the system or process remains within the validated state. This is done by monitoring the performance and output of the system or equipment. Furthermore, any changes to this system or equipment must be assessed and documented in order to ensure the product is safe and meets acceptance criteria.

Critical Aspects

Critical aspects of manufacturing systems include the functions, features, abilities, and performance or characteristics required for the manufacturing process and systems to ensure consistent product quality and patient safety. They should be identified and documented based on scientific product and process understanding.

Critical Quality Attribute, CQA (Critical-to-Quality)

A property or characteristic with specific nominal value and appropriate limit and range providing a particular quality attribute. A CQA typically is classed as a high-risk requirement, where the safety or efficacy of the product depends on the CQA being within the specified limits.

CCC (Mark)
The "China Compulsory Certificate"mark, commonly known as CCC Mark, is a safety mark for many products sold on the Chinese market. As of 2013, medical devices do not require this certification.

CDC

Center for Disease Control & Prevention (USA).

CDRH

Centre for Devices and Radiological Health (USA).

CE Marking

CE marking is a mandatory conformance mark on many products (including medical devices) placed on the single market in the European Economic Area. The CE marking certifies that a product has met EU consumer safety, health or environmental requirements. By affixing the CE marking to a product, the manufacturer declares that it meets EU safety, health and environmental requirements (CEN - Communité Européenne des Normes (European Committee for Standardisation).

Clinical Trial

Clinical trials are conducted to allow safety and efficacy data to be collected for health interventions (e.g. drugs, diagnostics, devices, therapy protocols). These trials can take place only after satisfactory information has been gathered on the quality of the non-clinical safety, and health authority/ethics committee approval is granted in the country where the trial is taking place.

Clinical Trial Sponsor

The clinical trial sponsor is responsible for the safety of subjects in a clinical trial and informs local site investigators of the true historical safety record of the drug, device or other medical treatment to be tested, and of any potential interactions of the study treatment(s) with already approved medical treatments.

Cleaning

Removal of contamination or soils from an item or surface to the extent necessary for its further processing and its intended subsequent use.

CMDCAS

Canadian Medical Devices Conformity Assessment System.

CMDR

Canadian Medical Device Regulation.

Conformity

Fulfilment of a requirement or meeting a requirement.

Conformity Assessment Body (CAB)

A body, other than a regulatory (competent) authority, engaged in determining whether the relevant requirements in technical regulations or standards are fulfilled.

CRO

A "Contract Research Organisation", also commonly known as a "Clinical Research Organisation", is a service organisation that provides support to the pharmaceutical and biotechnology industries. CROs offer clients a wide range of "outsourced" pharmaceutical research services to aid in the drug and medical device research and development process.

D

Data Integrity

Data integrity is the degree to which data is reliable and without error. Data must be accurate, attributable, contemporaneous, original, legible and available. A breach of data integrity occurs when any person manipulates or distorts data and submits the results of that data as valid.

Dead Leg

A dead leg in the world of piping terminology refers to an area of piping where there is insufficient flow or a tendency for water build-up or stagnation. The formal definition of a dead-leg states that pipelines for the transmission of purified water for manufacturing or final rinse should not have an unused portion greater in length than 6 diameters (6D rule) of the unused portion of pipe measured from the axis of the pipe in use.

Debugging

The process of locating, analysing and correcting suspected faults or machine issues.

Design Controls

Design controls are a collection of practices and procedures that are incorporated into the design and development process for a product such as a medical device. They provide a structure and clear path from user needs assessment to product delivery through a step-by-step process. Design controls ensure proper assessment of the design is completed during the design and development phase. Design controls are a requirement of quality systems such as 21 CFR Part 820 (medical devices), and for certain classes of devices and per ISO 13485 - Quality Management Systems.

Decommissioning

When a system is taken out of production service and stored in an adequate environment for potential future use.

Depyrogenation

A thermal process used to destroy or remove pyrogens (endotoxins). Typically, primary packaging components such as glass vials are subject to depyrogenation.

Detection Limit

The lowest amount of analyte in a sample that can be detected but not necessarily quantitated as an exact value for an individual analytical procedure. (Ref: ICH Q2.)

Design History File

The DHF is a repository for all of the documentation generated as a result of the design control process. The DHF serves as a complete record of the design.

Design Validation

Establishing by objective evidence that device or product specifications conform to user needs and intended use(s) defined in design documentation.

Debarment

The FDA has the authority to disqualify or remove researchers from conducting clinical testing of new drugs and devices when the agency determines that the researcher has repeatedly or deliberately not followed the rules intended to protect study subjects and ensure data integrity. Further, the FDA can disqualify a clinical investigator who has repeatedly or deliberately submitted false information to the agency or study sponsor in a required report.

Under its statutory debarment authority, the agency may also ban or "debar" from the drug industry individuals and companies convicted of certain felonies or misdemeanours related to drug products. Once individuals have been subjected to debarment, they may no longer work for anyone with an approved or pending drug product application at FDA. Debarred companies may no longer submit abbreviated drug applications.

Design Qualification (DQ)

The documented verification that the proposed design of the product is suitable for the intended purpose. DQs are typical deliverables for facilities, systems and equipment or processes.

Design Space

The multidimensional combination and interaction of input variables, e.g. material attributes and process parameters that have been demonstrated to provide assurance of quality. Working within the design space is not considered as a change.

Directives

Directives are legal requirements. These must be met by manufacturers. Standards such as ISO 13485 help companies meet the requirements of directives, such as "Guidelines Relating to the Application of the Council Directive 93/42/EEC on Medical Devices".

Direct Impact (System)

A system that is expected to have a direct impact on product quality. These systems are designed and commissioned in line with Good Engineering Practice (GEP), and, in addition, are subject to qualification and validation. Such systems include HVACs and clean utilities such as WFI (Water-for-Injection).

Diffusion Blending

A process in which particles are reoriented in relation to one another when they are placed in random motion and interparticular friction is reduced as a result of bed expansion (usually within a rotating container). This is also referred to as tumble blending.

Deviations

A deviation can be simply described as an unintended event which causes a test or verification to fail to meet expected acceptance criteria.

Degree of Invasiveness

A device, which in whole or in part, penetrates inside the body either through a body orifice or through the skin surface, is invasive. Invasiveness is generally categorised as invasive of a body orifice (including the surface of the eye), surgically invasive devices and implantable devices.

Device Master Record (DMR)

A compilation of records containing the procedures and specifications for a device. The contents of a DMR can contain local procedures such as SOPs and work instructions along with global or divisional specifications used to detail manufacturing processes, intermediate product or final product.

Drug Product

The dosage form in the final immediate packaging intended for marketing. The finished dosage form that contains a drug substance, generally, but not necessarily in association with other active or inactive ingredients. (FDA.)

Duration of Contact

In determining the classification of a device, the duration that the device is in continuous contact with the patient is defined as transient, short term or long term. The longer the device is in contact with the patient or user, the greater the risk. Therefore, this has to be taken into account when determining classification. Continuous use is defined in MEDDEV 2.4/1 as the uninterrupted actual use for the intended purpose. Where use of a device is discontinued in order that the device is immediately replaced with an identical device (e.g. replacement of a urethral catheter), this shall be considered as continuous use of the device.

E

Electronic Signatures

Electronic signatures are computer-generated character strings that count as the legal equivalent of a handwritten signature. The regulations for the use of electronic signatures are set out in 21 CFR Part 11 of the FDA. Each electronic signature must be assigned uniquely to one person and must not be used by any other person. It must be possible to confirm to the authorities that an electronic signature represents the legal equivalent of a handwritten signature. Electronic signatures can be biometrically based or the system can be set up without biometric features.

Encapsulation

The division of material into a hard gelatine capsule. Encapsulators should all have the following operating principles in common: rectification (orientation of the hard gelatine capsules), separation of capsule caps from bodies, dosing of fill material/formulation, rejoining of caps and bodies, and ejection of filled capsules.

Endotoxin

A pyrogenic product (e.g., lipopolysaccharide) present in the bacterial cell wall. Endotoxins can lead to reactions in patients receiving injections ranging from severe fever to death.

Equipment Qualification

Qualification means the process to demonstrate the ability to fulfil specified requirements. EQ consists of proving and documenting that equipment or ancillary systems are properly installed (Installation Qualification, IQ), work correctly (Operations Qualification OQ), and the different sub-systems work together as a system (Performance Qualification PQ) and actually lead to the expected results. Qualification is part of validation, but the individual qualification steps alone do not constitute a validated process.

Excipient

Substances other than the API which have been appropriately evaluated for safety and are intentionally included in a drug delivery system to provide a specific role in manufacturing, shelf-life or physical property.

Equipment Range

The full range that equipment is capable of performing, as per the manufacturer specification and tolerances. (A process may not utilise the full equipment range, operating over a narrower range.)

F

Factory Acceptance Testing (FAT)

An FAT or Factory Acceptance Test is an engineering activity that inspects and verifies that the equipment or system meets the requirements of the URS.

Failure Mode and Effects Analysis (FMEA)

A risk assessment tool that provides for an evaluation of potential failure modes and their likely effect on outcomes and/or product or process performance in order to prioritise risks and monitor the effectiveness of risk-control activities. It is often used to identify areas within a given process, product, or system that render it vulnerable.

FDA 483s

An FDA 483 letter typically includes a summary of findings and observations in relation to an audit or inspection where the FDA representatives have reason to believe GMP or other regulations have been violated or are not being met. In response to an FDA 483 letter, the company should address each item and provide a timeline for correction or request clarification of what changes are required.

Freeze Drying

Lyophilisation is the removal of ice or other frozen solvents from a material through the process of sublimation and the removal of bound water molecules through the process of desorption.

Functional Design Specification (FDS)

A functional design specification is a document that specifies how particular requirements are met – this can be a combination of how the equipment/process operates mechanically/automatically etc. An FDS is typically written in response to a URS

Fluid

A fluid is a substance that undergoes continuous deformation when subjected to a shearing force.

G

GAMP

Good Automated Manufacturing Practice (GAMP) is a set of guidelines for manufacturers and users of automated systems in regulated industries, specifically the medical device, pharmaceutical and biopharmaceutical industries. The application of GAMP and the validation of automated systems in manufacturing help ensure that regulated medical devices and medicinal products have the required quality and are manufactured according to good practices, meet regulatory and legal requirements and ensure patient safety.

Good Documentation Practices, GDP

The handling of written or pictorial information describing, defining, specifying and/or reporting of certifying activities, requirements, procedures or results in such a way as to ensure data integrity.

Granulation

A process of creating granules. The powder morphology is modified through the use of either a liquid that causes particles to bind through capillary forces or dry compaction forces.

Grade A Areas

Aseptic processing areas, critical in nature where sterile products are exposed to the environment, receiving no further sterilisation. High-risk operations (for example aseptic stopperage, filling, loading of the lyophiliser) occur in Grade A areas. They are considered ISO 5 under both dynamic and static conditions.

Grade B Areas

Aseptic processing areas where the sterile product is protected from the environment. Grade B processing areas are the background environments for Grade A areas and are considered ISO 7 environments in the dynamic state and ISO 5 environments under static conditions.

Grade C Areas

Non-critical areas where bulk product or materials are exposed to the environment, yet final sterilisation has not yet been performed. Grade C areas are support areas for non-sterile production activities; purification, formulation, and preparation of components, equipment, etc. for sterilisation. They are considered ISO 8 (Class 100,000) environments in the dynamic state and ISO 7 (Class 10,000) environments under static conditions.

Grade D Areas

Non-critical production areas, support areas, airlocks, or corridors. They are support areas for non-sterile production activities in closed systems; cell culture, or buffer and media preparation areas. Grade D airlocks are used for the movement of product, materials, and personnel into classified areas.

GHTF

Global Harmonisation Task Force

GxP

GxP is a general term for good practice with regard to quality guidelines and regulations. These guidelines are used in many fields, including the pharmaceutical, medical device and food industries. "X" is used as an umbrella letter representing different subjects or disciplines in industry. Some prime examples include GLP (Good Laboratory Practice), GDP (Good Documentation Practice), GEP (Good Engineering Practice) and GMP (Good Manufacturing Practices). Furthermore, the use of a lower case "c" as a prefix indicates "current" or "up-to-date".

H

Harm

Damage to health, including the damage that can occur from loss of product quality or availability.

High Level Risk Assessment (HLRA)

A high-level risk assessment that can be used at the beginning of a project to estimate the risk, such as the risks involved with bringing in new computerised/automated equipment.

HPLC

High Performance Liquid Chromatography (HPLC). An instrumental separation technique used to characterise or to determine the purity of a BDP by passing the product (or its component peptides or amino acids) in liquid form over a chromatographic column containing a solid support matrix. The mode of separation, i.e. reversed phase, ion exchange, gel filtration, or hydrophobic interaction, is determined by the column matrix and the mobile phase. Detection is usually by UV absorbance or by electrochemical means.

HVAC

Heating, ventilation and air-conditioning (HVAC) systems are used to control the environmental conditions within an area or manufacturing facility. HVAC systems also provide comfortable conditions for operators based in the manufacturing environment. Temperature, relative humidity (RH) and ventilation should not adversely affect the quality of products during their manufacture and storage, or the proper functioning of equipment.

Hydrogel

A biomaterial made up of a network of polymer chains that are highly absorbent and as flexible as natural tissue.

ICH

International Conference on Harmonisation of Technical Requirements for Registration of Pharmaceuticals for Human Use.

Intended Purpose

Intended purpose means the use for which the device is intended according to the data supplied by the manufacturer on the labelling, in the instructions and/or in promotional materials. (Chapter I section 1 of Annex IX of Directive 93/42/EEC.)

Immunoassay

A qualitative or quantitative assay technique based on the measure of interaction of high-affinity antibody with antigen used to identify and quantify proteins.

Impurity

Any component of the new active pharmaceutical ingredient which is not the chemical entity defined as the new active pharmaceutical ingredient OR any component present in the active pharmaceutical ingredient or final product which is not the desired product, a product-related substance, or excipient including buffer components.

Invasive Device

A device, which, in whole or in part, penetrates inside the body, either through a body orifice or through the surface of the body.

Ion Exchange Chromatography (IEC)

A gradient-driven separation based on the charge of the protein and its relative affinity for the chemical backbone of the column. Anion/cation exchange is commonly used for proteins.

IQ/OQ

Equipment IQ/OQ is defined as establishing documented evidence that all key aspects of the process equipment installation adhere to the manufacturer's approved specifications and any recommendations of the supplier of the equipment are suitably considered. The process/equipment must also operate as intended and all user requirements must be adequately fulfilled.

IFU

Instructions for Use.

Injunction (Plant)

An injunction is a judicial process initiated to stop or prevent violation of the law, such as to halt the flow of violative products in interstate commerce and to correct the conditions that caused the violation to occur. (FDA 21 U.S.C. 332; Rule 65, Rules of Civil Procedure.)

If a firm has a history of violations and has promised correction in the past but has not made the corrections, the injunction is more likely to succeed. However, the freshness of the evidence is critical. For an injunction action to be credible in the eyes of the Department of Justice (DOJ), the U.S. Attorney and the court, the evidence must be current. Timeliness is an important factor when considering an injunction action, with or without a Motion for Preliminary Injunction or a Temporary Restraining Order (TRO). However, case quality and credibility must not be sacrificed to meet guideline time frames.

The purpose of the guideline time frames is to limit, as much as can reasonably be expected, the need to update evidence. Updating entails extra work at all levels of the case development and review process and more importantly, delays obtaining an injunction which is intended to stop violations that adversely affect the safety or quality of products in commerce.

ISO

International Organisation for Standardisation. The agency responsible for developing international standards, e.g. ISO 13485 Medical Devices.

Isolator

A sealed enclosure, which provides full physical separation between the critical processing zone and the other surrounding processing zones. The internal surfaces of the isolator and of its contents are decontaminated, in accordance with defined objectives, by highly effective cycles, e.g. vaporised hydrogen peroxide. An enclosure capable of preventing ingress of contaminants by means of physical interior/exterior separation, and capable of being subject to reproducible interior bio-decontamination.

Isoelectric Precipitation

Isoelectric precipitation works by reducing the electrostatic forces to near zero, allowing the proteins to precipitate out.

ISO 13485

ISO 13485 is an ISO standard, published in 2003, that represents the requirements for a comprehensive management system for the design and manufacture of medical devices.

ISO 14971

An ISO standard, published in 2007 that provides a framework and requirements for a risk-management system for medical devices. This standard establishes the requirements for risk management to determine the safety of a medical device by the manufacturer during the product life cycle.

ISO 9001

ISO 9001 is an ISO standard that represents the requirements for quality management systems. It is used across industries and is not specific to medical devices like ISO 13485.

Item Master

The item master is a record of all components that a manufacturer buys, builds or assembles into its products. The item master includes information like the size, shape, material, manufacturer, manufacturer part number and vendor for each component.

IVD

In vitro diagnostic tests are medical devices intended to perform diagnoses from assays in a test tube, or more generally in a controlled environment outside a living organism.

IVDD

The In Vitro Diagnostic Device Directive delineates requirements that in vitro diagnostic devices must meet before they can be sold in the EU market.

Intermediate

A material produced during steps of the processing of an API that undergoes further molecular change(s) or purification before it becomes an API.

J

JIT (Just in Time)

A strategy used to monitor inventory levels with the goal of reducing inventory and associated carrying costs.

K

Kanban

A scheduling system that advises manufacturers what to produce, when to produce and how much to produce. Pioneered by Toyota, the approach is based on demand. Inventory is replenished only when visual cues like an empty bin, trolley or cart show that it's needed.

L

Limulus Amoebocyte Lysate Test (LAL)

A sensitive test for the presence of endotoxins using the ability of the endotoxin to cause a coagulation reaction in the blood of a horseshoe crab. The LAL test is easier, quicker, less costly and much more sensitive than the rabbit test, but it can detect only endotoxins and not all types of pyrogens and must therefore be thoroughly validated before being used to replace the USP Rabbit Pyrogen test. Various forms of the LAL test include a gel clot test, a colorimetric test, a chromogenic test, and a turbidimetric test.

Laminar Flow

Laminar flow is when fluid particles move in parallel layers, at a constant velocity.

Life Cycle (Validation)

The validation life cycle refers to the requirement to control and document all validation activities from the conception and URS stage to the retirement of equipment or a process. The life cycle approach ensures compliance throughout the life of the process/equipment while maintaining a validated state throughout the application of change control.

Linearity

The ability of an analytical procedure (within a given range) to obtain test results that are directly proportional to the concentration (amount) of analyte in the sample.

Line Clearance

The act of performing and documenting the removal of materials from a production or packaging line and cleaning prior to the introduction of a new batch or lot.

Lyophilisation (Freeze Drying)

Lyophilisation is the removal of ice or other frozen solvents from a material through the process of sublimation and the removal of bound water molecules through the process of desorption.

M

Mass Spectrometry

A technique useful in primary structure analysis by determining the molecular mass of peptides and small proteins. Often used with peptide mapping to identify variants in the peptide composition. Useful to locate disulfide bonds and to identify post-translational modifications.

Maximum Allowable Carry Over (MACO)

The amount of allowed product residue (carry-over) from lot-to-lot, batch-to-batch. This limit is based on the most conservative or lowest level of three MACO calculation methods (1) limited based on toxicity, (2) limit based on smallest therapeutic dose, and (3) worst-case dose.

Measurement Capability Index (MCI)

The Measurement Capability Index (MCI) represents the capability of the measurement system. It is used to evaluate the capability of the gauge to classify product against predetermined specifications.

Measurement System Analysis (MSA)

A study to determine the degree of error involved in measuring the given parameter. The measurement system involves the combination of operations, procedures, gauges, instruments, environmental conditions, people and software.

Medical Device

A medical device is an instrument, apparatus, implement, machine, contrivance, implant, in vitro reagent, or other similar or related article, including a component part, or accessory which is:

- recognised in the official National Formulary, or the United States Pharmacopeia, or any supplement to them,
- intended for use in the diagnosis of a disease or other conditions, or in the cure, mitigation, treatment, or prevention of diseases, in man or other animals, or
- intended to affect the structure or any function of the body of man or other animals, and which does not achieve any of its primary intended purposes through chemical action within or on the body of man or other animals and which is not dependent upon being metabolised for the achievement of any of its primary intended purposes.

Medicinal Drug Products (Finished Products)

Finished dosage forms (e.g. tablet, capsule, or solution) that contain the active pharmaceutical ingredient usually combined with inactive ingredients. Medicinal products are intended to furnish pharmacological activity or other direct effect in the diagnosis, cure, mitigation, treatment, or prevention of disease or to affect the structure and function of the body.

MDD

The Medical Device Directive is intended to harmonise the laws relating to medical devices within the European Union. Medical Device Directive 93/42/EEC was most recently reviewed and amended by 2007/47/EC.

MHRA

The Medicines and Healthcare Products Regulatory Agency (MHRA) is the UK government agency which is responsible for ensuring that medicines and medical devices work and are acceptably safe.

MSDS

Material Safety Data Sheet.

N

NCR

Non-Conformance Report.

NIH

US National Institutes of Health.

NOEL

No Observed Effect Level (in relation to cleaning validation).

Non-Conformity

A deficiency in a characteristic, product specification, CQA, process parameter, record, or procedure that renders the quality of a product unacceptable, indeterminate, or not according to specified requirements.

Non-Parametric Data

Where the type of data is non-variable. Also referred to as attribute data, e.g. visual inspection resulting in a PASS/FAIL result.

Notified Bodies

A notified body is a certification organisation which the national authority (the competent authority) of a member state designates to carry out one or more of the conformity assessment procedures or audits described in the annexes of the medical devices directives or GMP legislation.

NPI (New Product Introduction)

The market launch or commercialisation of a new product. NPI takes place at the end of a successful product development project.

O

Open System

An environment in which system access is not controlled by personnel who are responsible for the content of electronic records on the system (21 CFR, Part 11).

Outlier

A test result that is statistically different compared to a set of other test results obtained from the same sample or samples from the same lot of material.

Out-of-Specification

A recorded result that falls outside the established specification(s) or acceptance criteria.

Out-of-Trend

An analytical result which is within specification or acceptance criteria, but different from those usually obtained or expected. Out-of-trend results should be investigated by the same general principles as out-of-specification results.

Quantitation Limit

The lowest amount of analyte in a sample which can be quantitatively determined with suitable precision and accuracy for an analytical procedure. The quantitation limit is a parameter of quantitative assays for low levels of compounds in sample matrices and is used particularly for the determination of impurities and degradation products.

Overall Equipment Effectiveness (OEE)

A calculation for measuring the efficiency and effectiveness of a process, by breaking it down into three constituent components. (The OEE Factors — Availability x Performance x Quality.)

Overkill

A sterilisation process that is demonstrated as delivering at least a 12 Spore Log Reduction (SLR) to a biological indicator having a resistance equal to or greater than the bioburden level.

P

Pan Coating

The uniform deposition of coating material onto the surface of a solid dosage form while being translated via a rotating vessel.

Particle Count Test

This test covers verification of cleanliness. Dust particle counts are measured. The number of readings and positions of tests should be defined in accordance with ISO 14644-1 Annex B5.

Peptide Mapping

A technique which involves the breakdown of proteins into peptides using highly specific enzymes. The enzymes cleave the proteins at predictable and reproducible amino acid sites and the resultant peptides are separated via HPLC or electrophoresis. A sample peptide map is compared to a map done on a reference sample as a confirmational step in the identity profiling of a product. It is also used for confirmation of disulfide bonds, location of carbohydrate attachment, sequence analysis, and for identification of impurities and protein degradation.

Performance Indicators

Measurable values used to quantify quality objectives to reflect the performance of an organisation, process or system, also known as performance metrics in some regions. (ICH Q10.)

Performance Qualification (PQ)

Establishing by documented evidence that the process, under anticipated (controlled) conditions, consistently produces a product which meets predetermined requirements.

Precision

The level of agreement (scatter) between a series of measurements when a method is applied repeatedly to multiple samplings of a homogeneous sample or artificially prepared sample under the prescribed conditions. There are three types of precision; repeatability, intermediate precision and reproducibility.

Pressure Cascade

A process whereby air flows from one area, which is maintained at a higher pressure, to another area at a lower pressure.

Piping & Instrument Diagrams (P&IDs)

Engineering technical drawings that provide details of the connections and integration of equipment, services, material flows, plant controls and alarms. The P&ID also provide the reference for each tag or label used for identification.

PMA

Premarket approval by the FDA is the required process of scientific review to guarantee safety and effectiveness for Class III devices.

PMDA

The Pharmaceutical and Medical Devices Agency in Japan reviews applications for marketing approval of pharmaceuticals and medical devices. It also monitors their post-marketing safety and provides relief compensation for people who have suffered from adverse drug reactions from pharmaceuticals or infections from biological products.

PMS

Post-marketing surveillance is the practice of monitoring a pharmaceutical drug or device after it has been released on the market.

Process Design

Defining the commercial manufacturing process based on knowledge gained through development and scale-up activities.

Process Qualification

Confirming that the manufacturing process as designed is capable of reproducible commercial manufacturing.

Process Window

The selected operating range of machine settings/parameters that will produce product to meet all quality and product specifications.

Product Recovery

Product recovery is a critical and important step in the process. It is also referred to as "downstream processing". It is often the most expensive step in the process. For recombinant DNA-derived products, purification can often account for 90% of the total production cost.

Prospective Validation

Prospective validation is when validation is done in advance of commercial manufacturing.

Procedures

Also known as Standard Operating Procedures, or SOPs, these rules give directions for performing certain operations.

Protocols

Protocols give instructions for performing and recording certain discreet operations. Examples include engineering protocols, validation protocols etc.

Pure

A term typically used within pharmaceutical manufacturing, a product or substance is pure if it is free of contaminants, foreign matter, chemicals and harmful microbes.

Q

QMS

A QMS (Quality Management System) can be described as the organisational structure, procedures, processes and resources needed to implement quality management.

Quality

The degree to which a set of inherent properties of a product, system, or process fulfil requirements. (ICH Q9.)

Quality by Design

This is a systematic approach that begins with predefined objectives and emphasises product and process understanding and process control, based on sound science and engineering principles.

Quarantine

The status of materials isolated physically or by other effective means pending a decision on their subsequent approval or rejection.

Quality Policy

A document in which a company or organisation outlines its commitment and approach to quality. It usually sets out how the firm plans to achieve a high and consistent standard of quality. It should in some way speak to the customer or end user.

Qualification Plan

A Qualification Plan (QP) describes all the qualification measures and at which stage of the qualification the verification will be completed. It typically contains detailed descriptions of the necessary test measures and a description of the interdependencies of the individual tests. In some instances, there may not be a need or a requirement for a qualification plan. A validation plan can also serve to detail the qualification strategy.

QP

Companies that intend to manufacture or import medicinal products or intermediate products, for use in clinical trials or for market within the EU, must appoint a QP (Qualified Person), in order to comply with EU good manufacturing practice standards.

QPM

Quality Policy Manual.

QSP

Quality System Procedure.

QSR

Quality System Regulations.

R

Range

Range is defined as the interval between the upper and lower measurements required. The minimum specified range should be within the equipment range and validated to operate at all points within the range.

Recall

As defined by 21 CFR 7.3(g), recall means a firm's removal or correction of a marketed product that the Food and Drug Administration considers to be in violation of the laws it administers and against which the agency would initiate legal action, 2 21 CFR 806.2(h), e.g. seizure. Recall does not include a market withdrawal or a stock recovery. Recall does not include routine servicing. Recall also does not include an enhancement, as defined by this guidance.

Relative Humidity

The ratio of the actual water vapour pressure of the air to the saturated water vapour pressure of the air at the same temperature expressed as a percentage. More simply put, it is the ratio of the mass of moisture in the air, relative to the mass at 100% moisture saturation, at a given temperature.

Reusable Medical Device

A device intended for repeated use either on the same or different patients, with appropriate decontamination and other reprocessing prior to re-use.

Reusable Surgical Instrument

An instrument intended for surgical use by cutting, drilling, sawing, scratching, scraping, clamping, retracting, clipping or similar surgical procedures, without connection to any active medical device and which is intended by the manufacturer to be reused after appropriate procedures for cleaning and/or sterilisation have been carried out.

Re-Qualification

Requalification is designed to verify and ensure that the equipment/instrument/system is maintained in a qualified state after modification or after a stipulated time period (downtime).

Residual Risk

The risk level remaining after applying the identified controls on a high risk of harms and hazards manifestation.

Resolution

The smallest change in quantity that can be detected or provided by an instrument.

Residual Solvent

Organic volatile chemicals used or produced during the manufacture of APIs or excipients, or in the preparation of medicinal products.

Retain Samples

Samples that are kept for potential investigations and retests. It should be noted that retained samples are not a regulatory requirement, per Annex 10 or 21 CFR part 11.

Retrospective Validation

Retrospective validation is used for facilities or processes that have not completed formal validation. Historical data or a retrospective review can provide the evidence that the process or facility is operated as intended.

Rinse Sampling

Using a solvent to contact all surfaces of the sampled item to quantitatively remove target residue. The solvent can be water, water with pH adjusted, or organic solvent.

Right First Time

The Right First Time philosophy strives to create a culture of excellence. People are challenged with performing their tasks always in the correct manner to achieve the correct results always — right the first time.

Risk

The combination of the probability of occurrence of harm and the severity of that harm.

Risk Management

Risk management involves the systematic application of management policies, practices and procedures that identify, analyse, control and monitor risk. It is important to recognise that risk management should begin at the outset of the design and development phase of a project. The first step is to identify the user needs and intended use and application of the device.

RoHS

Restriction of Hazardous Substances in electrical and electronic equipment 2002/95/EC. An initiative that was adopted by the European Union (EU) in February 2003 and put into effect on July 1, 2006

Ruggedness

An indication of how resistant a test method or process is to typical variations in operation, such as those to be expected when using different analysts, different instruments and different reagent batches.

S

Scaffold

A structure of artificial or natural materials on which tissue is grown to mimic a biological process outside the body.

SKU (Stock Keeping Unit)

An SKU is a unique sales stock identifier.

Specifications

An approved document detailing the requirements with which the products or materials used or obtained during manufacture have to conform to. They serve as a basis for quality evaluation.

Specificity

The ability to assess unequivocally the analyte in the presence of components, which may be expected to be present.

Stability

Stability studies are used to demonstrate and justify assigned expiration or retest dates.

5S

5S is a Japanese methodology of organising and storing items in a work or lab environment. It has been adopted by many Western companies as a tool to help maintain standards and reduce errors and mix-ups. The "5s" represents each stage of the method:

Sort
Sorting out any items that are not in use and removing them to a more appropriate area such as a storage facilities or the bin.

Set-in-Order

The idea behind "set-in-order" is to be always organised. It requires "a place for everything and everything in its place". By setting things in order, we can help to make live processing and testing more efficient and reduce the risk of errors, omissions and accidents.

Shine

Regular cleaning is an important practice and it is always helpful to "clean as you go."

Standardise

Implement standard practices through SOPs and training. Standardisation can also be applied to workstation layout.

Sustain

Make it a habit! After implementing a 5s methodology, it is only effective if continuous efforts are made to "sustain" the changes.

Sterility Assurance (SAL)

The probability of a single viable microorganism occurring on an item after sterilisation. For a terminally sterilised medical device to be designated as "sterile", the minimum sterility assurance level must be SAL = 10-6 or better. When applying this quantitative value to assurance of sterility, an SAL of 10-7 has a lower value but provides a greater assurance of sterility than an SAL of 10-6.

T

Tableting

The reconstitution of a powder blend in which compression force is applied to form a single unit dose (tablet).

Tableting Press

Tablet press subclasses primarily are distinguished from one another by the method that the powder blend is delivered to the die cavity. Tablet presses can deliver powders without mechanical assistance (gravity), with mechanical assistance (automation), by rotational forces (centrifugal), and in two different locations where a tablet core is formed and subsequently an outer layer of coating material is applied (compression coating).

Traceability Matrix

A Traceability Matrix is a document that links the user requirements and specifications to where the verification and testing has been documented within the validation activities. A traceability matrix illustrates that all user requirements are traceable to the evidence-based test.

Turbulent Flow

Turbulent flow is when the movement of fluid particles are varying in velocity and direction.

U

Uniform

The product is manufactured consistently and will have the same quality between batches manufactured on different days.

UDI, Unique Device Identification

The UDI is a series of numeric or alphanumeric characters that is created through a globally accepted device identification and coding standard. It allows the unambiguous identification of a specific medical device on the market.

Uninterrupted Power Supply

An uninterruptible power supply (UPS) is a system for buffering the main power supply. If the power supply fails, the battery of the UPS supplies the required power. When the power supply returns, the UPS battery stops supplying power and is recharged.

Unit Operation

Unit operations are the individual steps in the process that modify materials and their properties at each step of the process. Each unit operation comes together to create a complete process.

User Requirement Specification, URS

The URS is a critical document that defines the requirements of a particular system, piece of equipment or process. Requirements such as the functional and operational aspects of the system are typically documented here.

USP

United States Pharmacopeia.

V

Validation

Validation is confirmation via documented evidence that the particular requirements for a specific intended use can be consistently fulfilled under anticipated conditions.

Validation Master Plan

A document providing information on a company's validation work programme. It typically details timescales for the validation work to be performed along with the key deliverables.

Verification

Verification means confirmation by examination and provision of objective evidence (i.e. documentation) that the specified requirements have been fulfilled.

Vaporised Hydrogen Peroxide (VHP)

Vaporisation of liquid hydrogen peroxide which results in a mixture of VHP and water vapour. The VHP mixture is used to decontaminate isolators.

W

Warning Letter

A warning letter is a piece of correspondence that notifies regulated industry about violations that the FDA has documented during its inspections or investigations.

WEEE Directive

Waste Electrical and Electronic Equipment Directive. European Community Directive 2002/96/EC where manufacturers are responsible for disposing of electrical/electronic waste.

WFI (Water-for-Injection)

WFI is sterile and pyrogen-free water containing no less than 10 CFU/100ml (Colony Forming Units) with a sample size of between 100 and 300 ml and an endotoxin level of < 0.25 EU/ml.

WHO

World Health Organisation.

WI

Work Instructions.

Witnessed By

When signed or initialled, this constitutes legal proof that the individual signing is physically present and observes the step, calculation, or operation being performed by someone else, and that all entries of data are true and accurate.

Worst Case

A set of conditions or parameters which, in combination with product specification or attributes at their limits, pose the greatest challenge to the process.

X

--

Y

--

Z

Zone Classification

Zone classification refers to GMP areas which include controlled (aka classified) and non-controlled manufacturing areas. Areas may be classified based on EU Grades A–D and/or ISO Class 5–8 (in the US - Class 100–Class 100,000 areas.

www.ingramcontent.com/pod-product-compliance
Lightning Source LLC
Chambersburg PA
CBHW050213230526
45470CB00001B/367